Firepower

History of the Aircraft Gun

Scott Vadnais & Bill Holder

Schiffer Military History
Atglen, PA

Acknowledgments

David Menard, Air Force Museum Research Department
Wright Patterson Air Force Base, Ohio

Helen Kavanaugh, Office of Public Affairs
Aeronautical Systems Center
Wright Patterson Air Force Base, Ohio

Book Design by Ian Robertson.

Library of Congress Catalog Number: 97-81448

Printed in China.
ISBN: 0-7643-0726-6

We are interested in hearing from authors with book ideas on related topics.

Published by Schiffer Publishing Ltd.
4880 Lower Valley Road
Atglen, PA 19310
Phone: (610) 593-1777
FAX: (610) 593-2002
E-mail: Schifferbk@aol.com.
Visit our web site at: www.schifferbooks.com
Please write for a free catalog.
This book may be purchased from the publisher.
Please include $3.95 postage.
Try your bookstore first.

In Europe, Schiffer books are distributed by:
Bushwood Books
6 Marksbury Road
Kew Gardens
Surrey TW9 4JF
England
Phone: 44 (0)181 392-8585
FAX: 44 (0)181 392-9876
E-mail: Bushwd@aol.com.

Try your bookstore first.

Table of Contents

Introduction

In the early days of aircraft use during World War I, so the story goes, scarf-draped pilots waved at their adversaries from their multi-wing creations as they passed each other during observation missions.

But it quickly became evident that these magnificent flying machines could also serve as airborne weapon platforms providing a visual perspective never available before.

Then, again as the tale tells, political correctness went away with the first hint of aerial combat being born. Reportedly these 'Knights of the Air' actually fired their hand-held weapons at each other.

The first operational use of the aircraft gun in the swirling skies of World War I. The conflict set the scene for the aircraft gun which would carry to the present time. (Air Force Museum Photo)

Respect was still maintained, though, when a pilot from one side was killed in action, a plane from the other side would oft-times make a tribute pass honoring the fallen gladiator.

But the chivalry quickly faded when permanent gun mounts were installed, specifically for the purpose of removing enemy aircraft from the sky. Initially, there would be forward-firing fuselage-mounted guns which required the pilot to point his plane directly toward the enemy plane. It was soon realized what a great ground attack weapon the airborne gun had become.

In addition to the attack capabilities of the early aircraft gun, it soon became evident that the aircraft gun could also be used to protect the aircraft upon which it was mounted. Protection of the aircraft from the rear was one of the early priorities of these guns.

Unlike the fixed front-firing guns, many of these early protective guns would be manned by a gunner who could maneuver the gun on its mount and cover a wide arc of sky. Later, ring mounts would provide a sounder and more flexible mounting base for the guns.

Even though early aerial combat was a deadly game, it was a game to many pilots who became famous for their kill scores. You'd really have to call them the first of the 1990s style of TOP GUNs.

As time went by, bombers would also start to acquire guns in some cases-many guns, except in this application the main purpose of guns was for protection. It's continued that way to today.

Initially, aircraft guns were of the machine gun variety but later the more-explosive punch of small rapid-fire cannons would be carried aloft.

With the advent of World War II, fighters and bombers bristled with guns. Forward firepower was unbelievable from certain fighters and light bombers. In some cases, there were field modifications for the building of "Gunships" with added forward-firing guns for mauling a target.

With fighters, most of the firepower was installed in the leading edges of the wings. Most bomber gun locations were

By the end of the war, the aircraft gun had become an effective aerial weapon system. Here is a standard mount arrangement, shown here on a Spad VIII C.1. (Vadnais Photo)

The B-26K was possibly the ultimate gun-equipped machine carrying eight .50-cal machine guns in the nose. (Vadnais Photo)

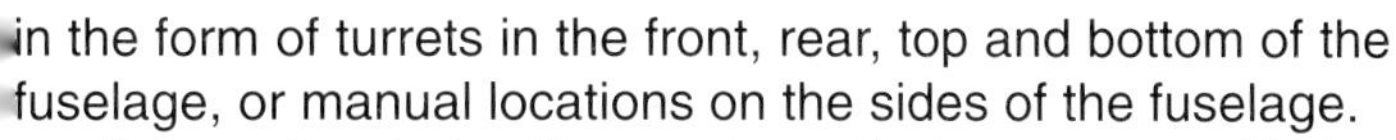

in the form of turrets in the front, rear, top and bottom of the fuselage, or manual locations on the sides of the fuselage.

It was also during the great war that guns started being supplemented by rockets for ground attack purposes.

There were even bombers modified with a small arsenal of guns which were to serve as protective escort planes flying along with a bombing formation.

With the advent of the B-29, the gunner was no longer pressing the trigger on his gun, but instead it was being done remotely from another location in the aircraft.

During the Korean War, the actual start of the combat jet age, aircraft guns still played heavily, especially with the F-86 Sabre Jet in MiG Alley confrontations.

During the 1960s, aircraft guns would become a part of the helicopter mission. These choppers would play an important role in the Vietnam War.

The B-52 would be the bomber of the century, but interestingly, there would be only one gun. It was a potent tail stinger that accomplished two kills in Southeast Asia. The following two bombers, the B-1 and B-2, possibly the final manned bombers, have no guns onboard. A majority of the Vietnam kills, though, would come from air-to-air missiles instead of guns.

Even with some trends indicating the need for guns having been reduced, there were also aircraft designed specifically around a gun. The three famous gunships, using the C-

With eight Colt-Browning M-2 machine guns in the wing leading edges, the P-47 Thunderbolt was an awesome air-to-air and air-to-ground weapon system. (Air Force Museum Photo)

With the advent of the Korean War came the swept-wing F-86 Sabre Jet. All its 'gun power' was carried in the forward fuselage. This A version carries six M-3 .50-cal machine guns which really got the job done in Mig Alley. (Bob Shenberger Photo)

47, C-119 and the ultimate C-130s, firing sideways out of the fuselage proved their worth many times over during the Vietnam conflict. The role was strictly ground attack.

Later would come the A-10 Warthog which carried a huge Gatling gun in its fuselage for close-in air support. It too proved itself in combat, Desert Storm in this case. There has also been a close air support (CAS) version of the F-16, whose mission was designed around an under-fuselage 30-mm gun pod.

With the advent of new air-to-air and air-to-ground missiles, there were those who thought that the day of the aircraft gun was over. In fact, the F-4 Phantom's early versions carried no guns. Experience showed, though, that they were still needed and later versions saw them returned.

The latest versions of USAF and US Navy aircraft, i.e. the F-15, F-16, and F/A-18, all carry rapid-fire Gatling Guns. But then there is the famous F-117 stealth fighter which carries no gun, actually no kind of protection weapons at all. Interesting that the F-117 got an "F" designation in the first place.

For the latest USAF fighter, the F-22, there was discussion on whether a gun would be needed. It was decided, though, that the old reliable gun system would be retained.

The Air Force actually has long held a aircraft gun competition, called William Tell, which measures the Top Gun pilots from a number of USAF and Canadian units.

The helicopter became a gun hauler with great effectiveness in the Vietnam War. Here, a UH-1D has a side-mounted .50-cal machine gun in the side door position. (USAF Photo)

Needless to say, the whole history of the use of the airborne gun has been interesting. It's a story that really has never been told. It's the story of a weapon that has come and gone, and come back again, depending on changing warfare scenarios through the decades which started early in this century.

And if history holds true, it will continue to be used in ever-different applications into the next century. Old habits die hard.

The days of many guns have vanished with modern fighters. The F-15 Eagle is a good representative of the trend carrying a single 20-mm M-61 multi-barrel gun. Also lots of missiles, but still a gun in position. (McDonnell Douglas Photo)

Chapter 1:
Early Aircraft Gun History

At first it was bombs as the prime mission of the early military aircraft. Like the guns that would shortly follow, these early bombs were crude and inaccurate.

But once the development of aircraft guns had started, the advancement was remarkable. But when you think about it, the time period was during World War I, and the pressure to develop advanced aircraft weaponry was immense.

It's amazing to learn that the French had even developed a 37-mm cannon which was experimentally fired through the propeller. The weapon proved to be effective on a limited basis, but having a solid place to firmly mount the gun proved to be a troublesome problem. It was necessary to make a direct hit the first time since the recoil from the weapon would rock the weapon into a new position destroying the aim for the next shot.

It would come to pass that machine guns, because of their rate of fire, would prove to be the weapon of choice. The weight limitation of these fragile aircraft limited the amount of ammunition that could be carried aloft along with the size of the weapon.

But possibly more important was the fact that there was NO armament to protect against an enemy attack. It's actu-

Machine guns on early WWI era aircraft normally were mounted on the top of the fuselage directly in front of the pilot. (Holder photo)

Sighting was done with early aircraft guns by primitive sights such as the one shown mounted on these early fighters. (Air Force Museum Photo)

Although the guns on this WWI fighter are mockups on a restored example, the mounting technique can be clearly seen. (Holder Photo)

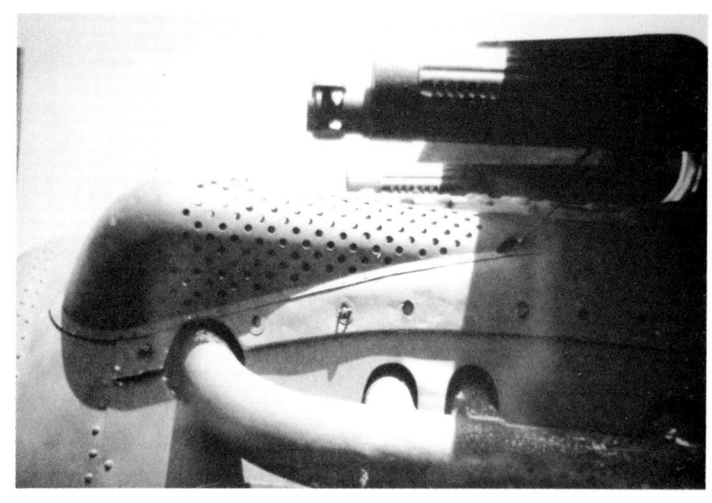

Looking up from under the front fuselage of a Spad VII, the fuselage-mounted .303-in machine guns is visible. (Air Force Museum Photo)

A near head-on view of the Spad VII .303 machine gun installation, a view that meant destruction for many enemy aircraft. (Vadnais Photo)

The Spad VIII C.1 sported a pair of nose-mounted machine guns which, of course, both had to fire through the spinning prop. (Vadnais Photo)

One of the most famous fighters of World War I, the Sopwith Camel carried a pair of Vickers .303 machine guns. (Vadnais Photo)

Preparing for another combat mission against the Axis, this Nieuport 28 WWI fighter sports a pair of .303 Vickers machine guns. In the AEF versions, the machine guns were of the Marlin brand. (Air Force Museum Photo)

This French version of a Nieuport 28 backgrounds a proud French pilot. The forward fuselage-mounted machine guns are clearly visible. (US Air Force Photo)

ally reported that some pilots took along stove lids to sit on to protect against attacks from underneath.

All the early land-use guns that were modified for the new aerial use. Changes were made to the guns to increase the firing rates since time within range of the target was limited.

There were two types of aircraft guns during these early days, not really much different from the types that would follow in the decades to come.

First, there were the fixed guns that mounted solidly on the aircraft fuselage. Of course, with this type of mounting it was necessary to maneuver the aircraft such that the gun would be pointing toward the potential target. It required excellent pilot skill. Also, the pilot was the gunner in this situation which required him to be quite coordinated.

With the solid type of mounting, it was often necessary to fire through the whirling propeller directly in front. Early French efforts were not that successful with a small percentage of the bullets impaling the propeller. Definitely not a good situation!

It would not be until later that a synchronizing gear set-up enabled the shots to be spaced between the blades.

There was one German Fokker innovation that designed a device that shut down the firing mechanism as the blade was passing through the field of fire, and then starting it up again when the coast was clear. It would be a Rumanian engineer, Georges Constantinesco that built an effective synchronizing mechanism that was based on a hydraulic system that was the probably the best technique of the many tried.

There were also several attempts made in the United States, by A. L. Nelson and Glen D. Angle, whose efforts developed the standard US synchronization system.

The German Albatross D.111 was one of the more effective firing-through-the-prop versions. A pair of Spandau 7.92-mm machine guns were mounted on the fuselage. Since the

The tri-winged version of the Dr.1 was similarly armed with the Spandau machine guns. (Vadnais Photo)

Note the machine gun mounting position which is much further back from the propeller. The open engine compartment is located directly in front of the gun position. (Vadnais Photo)

Testing was at a high level of activity in the United States. Shown here ia a synchronization test which took place at McCook Field near Dayton, Ohio. (Air Force Museum Photo)

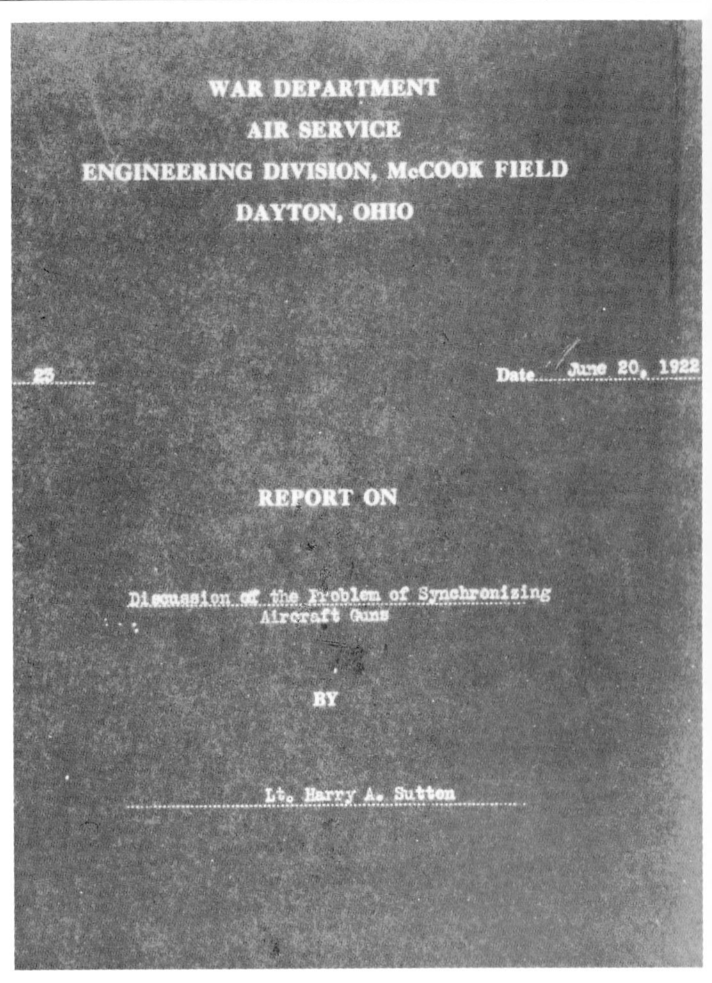

WAR DEPARTMENT
AIR SERVICE
ENGINEERING DIVISION, McCOOK FIELD
DAYTON, OHIO

25 Date June 20, 1922

REPORT ON

Discussion of the Problem of Synchronizing Aircraft Guns

BY

Lt. Harry A. Sutton

A number of technical reports were published on results of synchronization testing. Here is the cover of one of them. (Air Force Museum Photo)

A primitive (by today's standards) test of a Browning .30-cal machine gun at Kelly Field, Texas in 1924. (Air Force Museum Photo)

An F2-G gun is equipped with a Type B-3 Synchronizer at the end of a 100 hour test. (Air Force Museum Photo)

An advanced gun mount was the Browning Double Flex Gun Mount, shown here at Kelly Field, Texas. (Air Force Museum Photo)

The early fighter pilot had an added load with the sight and machine gun at his immediate front. (Air Force Museum Photo)

One of the mainstays of early aircraft guns was the dependable Lewis .30-cal machine gun. (Holder Photo)

top of the bi-wing aircraft's Mercedes engine protruded above the top of the cowling, the guns had to be spread, firing around the engine.

But mention WWI fighters, and the several versions of French Spads quickly come to mind. Thousands of this low-upper-wing fighter were built with one- and two-Vickers machine gun installations firing through the prop.

But early pilots quickly learned that the immediate threat wasn't always directly in front of them. Enemy threats could come from the sides, bottom and rear of their aircraft. As a result, consideration was made on guns that could fire in those directions and not requiring the aircraft to be pointed in those directions. Hence, the development of free guns.

A number of different mounting techniques were investigated with firing locations further back on the fuselage along with forward fuselage positions when a multi-engine aircraft was being used.

The new gun locations, of course, required a specific gunner in position to physically move the guns into firing position.

The Lewis and Browning machine guns were the weapons of choice for allied aircraft. The guns were made much

How's this for a nifty gun mount equipped with a ratchet. The French Packard LePere Lusac II aircraft carried a pair of both .30-cal Marlin and Lewis machine guns. (Vadnais Photo)

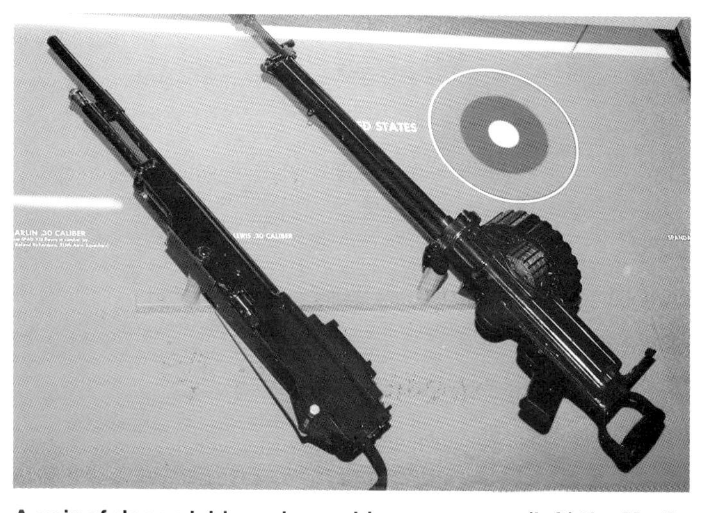

A pair of dependable early machine guns were (left) the Marlin .30 cal and the Lewis .30-cal guns. (Vadnais Photo)

Joe Gertler, a WWI gun replicator, has constructed at least 200 WWI replicas for collectors and plane restorations. (Holder Photo)

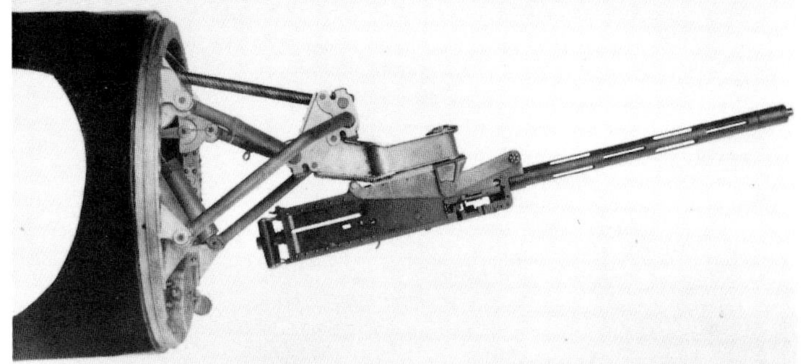

The effectiveness of the aircraft gun was greatly increased with the ring mount. This is the Type E-1 version. (Air Force Museum Photo)

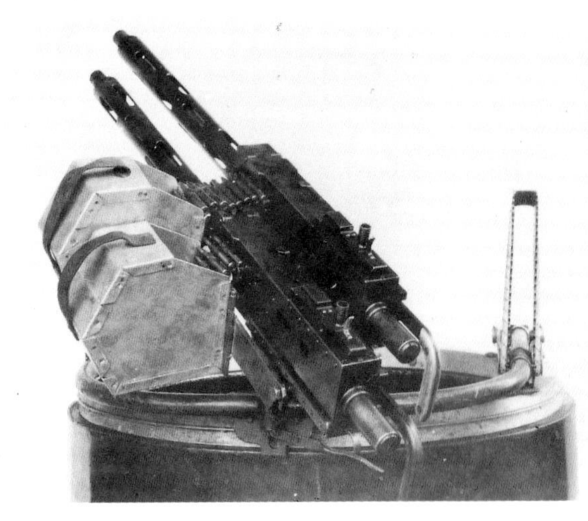

Another ring-mount set-up is shown with this twin machine installation. (Air Force Museum Photo)

more effective with the use of flexible ring mounts which allowed the guns to be more quickly and smoothly moved to firing location. Some of the free guns were equipped with primitive automatic wind compensators.

These early ring mounts were extremely flexible enabling the gunner the capability to cover wide ranges of fire including even almost straight down.

Also in use during the period was the so-called pedestal mount system where the gun was mounted on a fixed platform, and as such, was not as flexible as the ring mount system. At the end of the war, guns as large as .50-cal Browning machine guns were in use.

The highly-successful ring mount concept was used by the British Handley Page 0/100 and 0/400 twin engine bombers. The types used a front-fuselage-mounted ring mount for a pair of .303 Lewis machine guns. The location was far out on the nose with the machine gun barrels reaching far forward of the nose.

It's interesting to note that many of these same concepts were utilized in aircraft during the early World War II time period.

Following WWI, there was some feeling in the general populous that the last great war had just ended and it was not necessary to develop any additional aircraft-mounted weapons. Fortunately, leaders at the time saw it differently and the Boeing Company developed the so-called GA-2 ground-attack aircraft.

The Curtis A-3B carried a ring mounted machine gun in an installation directly behind the pilot. (Air Force Museum Photo)

One of the early attempts of bomber front end protection is demonstrated by the Caproni CA 36 bomber which shows a pair of Revelli 6.5-mm machine on a ring mount. (Vadnais Photo)

Note the interesting ammunition feed mechanism on this period machine gun. (Air Force Museum Photo)

The details of this ring mount are clearly evident in this photo of an American Dayton-Wright DH-4. The set-up supports a pair of Lewis .30-cal machine guns. (USAF Photo)

Although, the term "Gunship" is used today as a modern term, this early 1920s aircraft would certainly have to fall under that descriptor. With the success of ground strafing enjoyed during the war, it was felt that this heavily armed machine would optimize that capability. And just looking at its armament, it would be hard to argue that conclusion.

When you hear how much armament and guns the plane carried, it's hard to believe that it was just a single-engine plane. But consider that the monster engine displaced 2700 cubic inches with l8 cylinders, the capability to tote the heavy load was there.

The firepower of the GA-2 was awesome! Consider it carried a 37-mm cannon, .50-cal machine guns, a brace of .30-cal guns. The cannon was located just above the landing gear while the pair of .50-cal machine guns directly above the cannon had a downward seep to 60 degrees from vertical and l5 degrees from the horizontal. The other .50-cal gun was carried in a rear gunner in a fixed position, and with an advanced technique, was fired remotely by either the top rear gunner or by the lower gunner. Scarff-mounted Lewis guns completed the impressive array.

Probably the most recognized pilot of the World War I era was the incomparable Captain Eddie Rickenbacker standing next to his famous Spad fighter. (USAF Photo)

A view of members of the famous 94th Aero Squadron during the time when the organization switched from Nieuport 28s to Spad 13s. (USAF Photo)

Chapter 2: World War II Guns and Cannons

Through the 1920s and 1930s, with the pressure of war not pushing advanced developments, the development of new aircraft gun technology remained relatively static.

Browning and Lewis would be the prime US manufacturers of machine guns while the American Armament Company was the prime producer of cannons when the seeds of World War II were being sown in the late 1930s. Madsen was the prime name for the British with the manufacture of both machine guns and cannons. During this period, there was increased interest in the larger caliber, heavier-hitting cannons, but still the standard machine gun would play a heavy role for years to come.

As was the case with the initial guns from World War I, there would be the same fixed and flexible mounting techniques.

The fixed gun technique, though, would be applied much more widely than just the fuselage mountings of the WWI aircraft. Granted, the fuselage mounting was still in use, but with fighter application, the prominent location was within the wings, as many as four on each wing as illustrated by the P-47 Thunderbolt. Generally, with the wing installation, the fir-

One of the P-35's pair of machine guns is visible from this right side view. (Vadnais Photo)

The P-35 was a transitional aircraft developed between the wars that would help set the standard for World War II fighters. (Air Force Museum Photo)

This head-on shot of the P-35's nose shows its pair of machine guns, one a .30- and one a .50-caliber. (Air Force Museum Photo)

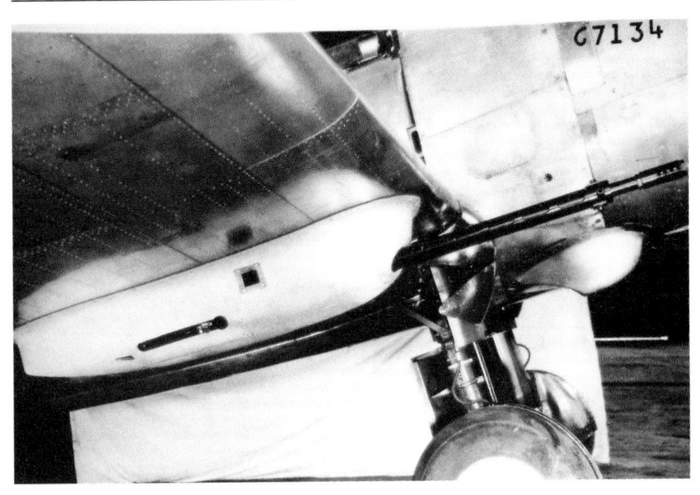

The P-36F, a later version of many earlier models, sported a 23-mm gun mounted under a cover underneath the wing. Other versions had many different combinations of guns. (Air Force Museum Photo)

One of the most awesome firepower fighters was the P-38 Lightning, shown here on a firing range during World War II. (Air Force Museum Photo)

At an unidentified P-38 base, this Lightning gets a reload for its array of nose guns. (Air Force Museum Photo)

No automatic loading system for the P-38. It took lots of muscle to do this job. (Air Force Museum Photo)

The P-38E was the first model to be coined 'Lightning' and adapted the Hispano gun as a main weapon. (Air Force Museum Photo)

The four nose guns of this P-38J provide the Lightning with tremendous forward firepower. (Air Force Museum Photo)

Bigtime firepower sits on the nose of the P-38L with four .50-caliber machine guns and a single 20-mm cannon. That will definitely get you attention! (Vadnais Photo)

Note the recessed guns in the nose of this later L model which are arranged symmetrically around the rear of the nose cone. (Air Force Museum Photo)

ing was accomplished through a remote-control solenoid attached to the gun receiver.

The fixed gun concept would also be used in some very interesting non-factory set-ups. There were some very interesting modifications with the multi-gun forward firing arrangements being the most interesting. On several occasions, as many as four guns were installed in the noses of light bombers in order to provide awesome firepower for ground attack.

The flexible gun systems found more application with the larger bomber aircraft with a number of the installations being open to the brutal high-altitude winds. For example, the A-20A light bomber had an open-air rear flexible gun location in the rear cockpit.

Probably most familiar were the flexible gun positions on the heavy bombers of the period. Both had open air flexible gun positions in the waist positions, along with flexible turret locations on the upper and lower-fuselage locations, along with standard flexible gun mounts in the tail and nose.

During the later days of the war, turret advances included the incorporation of remotely-controlled units where no on-site gunner was required. The B-29 heavy bomber was a prime example of that concept.

No matter which type of installation a gunner manned in those days, there was one rule that applied in both situations. It was best to fire in short bursts since the enemy aircraft would be within range for only a short period.

Forward firepower was of prime importance for the ground attack mission, and a number of interesting modifications were adapted through the war years, some of them coming in the field and others in the factory. Several light bombers had their noses bristling with machine guns for the ultimate 'hose-down' of an enemy area.

Protection was also a prime use of the aircraft gun during World War II. One experiment, although not totally successful involved the modification of B-17s to flying arsenals to aid in protecting the bomber fleets in missions over Europe. It

This exposed firepower is the 37-mm cannon and 30 rounds of ammunition which is carried by the P-39Q, the final model of the Airacobra. (Vadnais Photo)

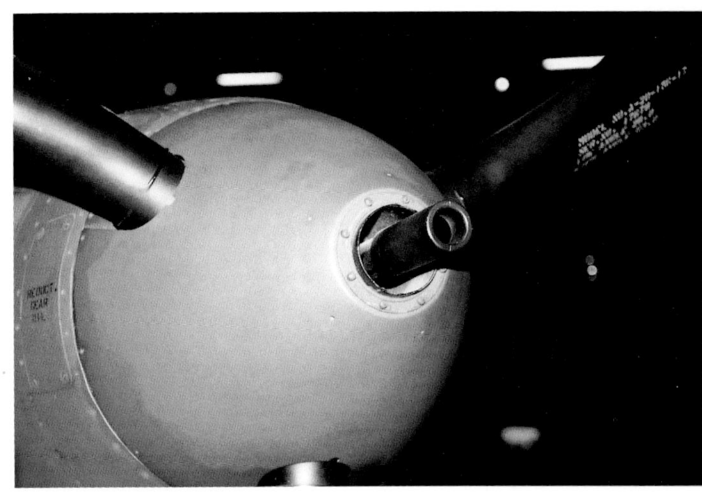

A view of the business end of the P-39Q's 37-mm cannon. But there were also other machine guns mounted in the nose and under the wings. (Vadnais Photo)

was at a time when escorting fighters weren't able to accompany the bombers all the way to target.

Initially, guns were the only offensive appendage to the single engine fighters and attack aircraft, but as the war went along, that all changed. In addition to the guns, wing-mounted rockets were coming into play to augment the gun firepower. During the era, the technology had not advanced to the point of guided rockets or missiles, that would come later.

Following is a review of the magnificent single engine fighter and attack aircraft and multi-engine bomber machine gun and cannon systems, Allied, Axis, and Japanese, that emerged during the great conflict.

FIGHTERS & ATTACK AIRCRAFT

Lockheed P-38 Lightning (USA)

It was known as the Lightning, and with its in-excess-of 400 mph capability, the name was appropriate. And with its weapon suite, it was doubly apt.

On the center body section, there was mounted a 23-mm Madsen cannon and four .50-cal Browning machine guns firing straight ahead of the cockpit. Other versions of the potent single-seat fighter had varying weapons combinations. Early models carried a 37-mm Oldsmobile cannon, two .5-inchers and a pair of Colt .3's. Another combination had four .5-in machine guns and the same 37-mm cannon. There was also the incorporation of the RAF Hispano gun on the E model.

Almost ten thousand Lightnings were produced in a number of different mission configurations.

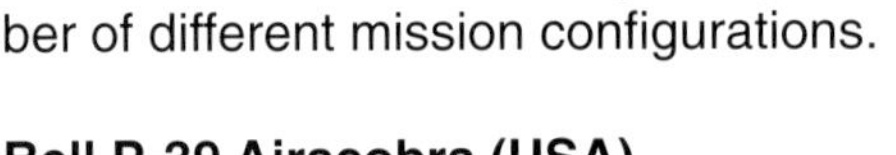

Bell P-39 Airacobra (USA)

Developed in the late 1930s, the Army P-39 provided significant contributions throughout the war. The model was characterized by the propeller hub-mounted 37-mm cannon with a round store of 30 rounds. But there was more.

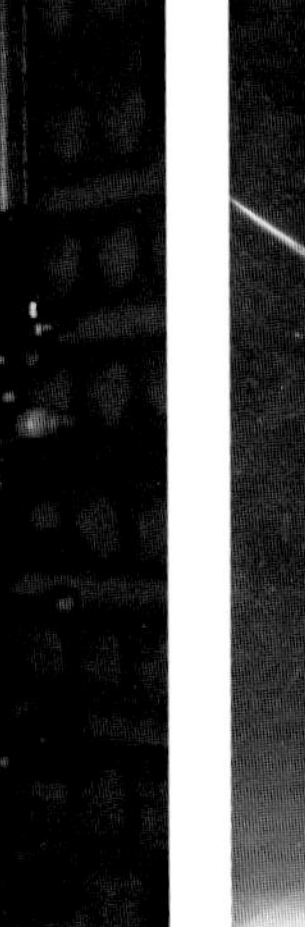

Note the rotating round holder for the P-39's 37-mm cannon which fed the weapon which fired through the nose cone. (Vadnais Photo)

This P-39 lights up the sky on a test range with the 37-mm cannon, and the array of .50-cal machine guns all doing their thing with tracer rounds. (Air Force Museum)

The follow-on to the P-36, the P-40 Warhawk carried all its firepower on the wing leading edges. This P-40E shows three of its six .50-cal machine guns. (Vadnais Photo)

An excellent view from the underneath of a P-40 wing shows the details of the gun installation. (Air Force Museum Photo)

Dramatic tracer round testing from this P-40. (Air Force Museum Photos)

This P-40E is out on the prowl looking for enemy aircraft. (Air Force Museum Photo)

The A-36 was a ground attack version of the P-40 and was well armed for the dangerous mission. There were six forward-firing .50-cal machine guns, two of which are shown here mounted on the lower forward fuselage. (Vadnais Photo)

One of the early D versions of the P-47 carried eight .50-cal machine guns, as did all the Thunderbolts. This version also incorporated front and rear armor protection and bullet proof glass against enemy firepower. (Air Force Museum Photo)

Note the varying lengths of the four Colt-Browning M2 machine guns, only half the total number of guns for the P-47 Thunderbolt. Also note the tremendous amount of ammo carried within the wing volume. (Air Force Museum Photo)

The C version would have four machine guns synchronized to fire through the prop. The British used the plane although that version used a 20-mm cannon in the nose, two machine guns on the outside fuselage, and four more guns in the wings.

Curtis P-40 Warhawk (USA)

This famous model evolved under the P-36 designation which featured one .50- and .30-cal. Browning machine gun mounted above the radial powerplant. There would then be a number of different armament configurations with guns added in the wings.

The follow-on P-40, made famous in the early China conflict, would incorporate the new Allison engine which enabled the shortening of the nose. Along with the engine change, the use of fuselage-mounted guns was discarded with six .50-cal. guns being added on the wings. As the performance increased, the P-40 became a formidable system with its tremendous firepower.

Republic P-47 Thunderbolt (USA)

The Thunderbolt was a brute of a plane grossing out at over ten tons fully loaded. Included in that massive bulk was tremendous firepower with eight Browning machine guns carrying 267, 350, or 425 rounds. The barrels were in an interesting configuration, with the inner barrel stretching out the longest and the outboard barrel flushed up with the leading edge of the wing.

The plane was one of the most popular of the war, and many air forces used the model well into the 1950s.

North American P-51 Mustang (USA)

Considered by many to be the best World War II fighter, the P-51 Mustang appeared in many forms, actually first flying

In addition to the potent machine guns, P-47s could also carry racks of rockets which were attached to the wing tank adapters. (Air Force Museum Photo)

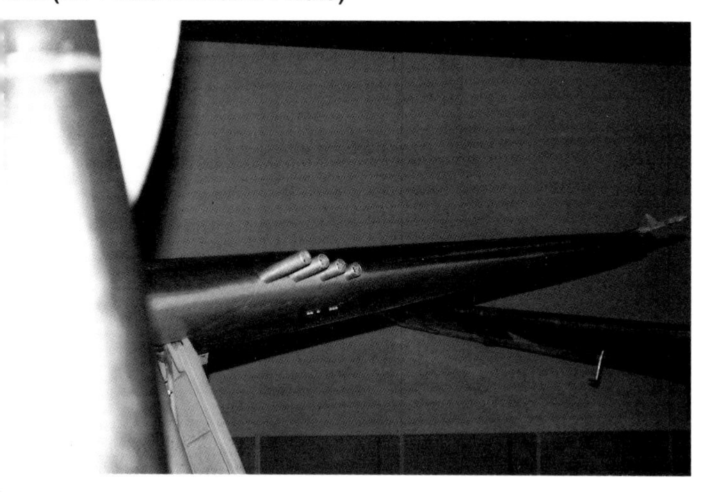

The Air Force Museum has an excellent example of the Thunderbolt breed on display. Here's four of its eight .50-cal machine guns looking at you. (Vadnais Photo)

Female production workers are shown working on a P-51's twin, 50-cal guns. (USAF Photo)

Ask any World War II buff which is the best fighter of that era, and many will tell you that it's this P-47. The plane was a fine combination of performance, durability, and firepower. (Air Force Museum Photo)

operationally for the Royal Air Force. In that configuration, the heavily-armed model carried four .303 guns in the wings, a pair of .5-inch guns in the wings, and two more .50-cal. machine guns in the nose.

The first AAF versions, the P-51A and P-51B, carried guns only in the wings, four .50-cal. weapons to be exact. The ground attack version of the Mustang(the A-36) was armed like an armory with six .50-cal guns in the wings. Later versions of the Mustang were similarly armed, with six .50-cal Browning MG53-2 machine guns with 270 or 400 rounds each. The later F-82 Twin Mustang still carried the same six machine guns armament suite even though it was a much larger aircraft.

Bell P-59 Airacomet (USA)

Although the jet-powered P-59 was not around for air combat in World War II, some versions of the fighter still carried armament. It was a similar set-up to the earlier P-39 which also featured nose guns, but with the P-59, it usually consisted of a 37-mm cannon and three .50-cal machine guns.

Northrop P-61 Black Widow (USA)

Featuring a menacing cold black paint scheme, the Black Widow had a complement of firepower to back it up. Carried in the belly were four potent forward-firing 20-mm M-2 Cannons. But with the final 250 B and C versions, the firepower was enhanced significantly with the addition of a dorsal turret which was located directly behind the cockpit.

No gunner up there, as the four .50-cal guns were controlled remotely from the front or rear sight station, but actu-

The four .50-caliber machine guns greatly protrude from the leading edge of this P-51A. (Air Force Museum)

This highly decorated P-51B is armed with four wing-mounted .50-cal machine guns. (Air Force Museum Photo)

ally fired by the pilot. It was a development that would later be used in the B-29 Superfortress.

Curtis SB2C/A-25 Helldiver (USA)

Although mostly known as a ship attack aircraft, the Navy Curtis Helldiver was also loaded for bear with its gun complement. First, there were either two 20-mm cannons or four .50-cal machine guns in the wings, but there was also rear protection in the form of a rear cockpit where either a pair of .30-cal or a single .50-cal gun were handled manually by the rear gunner.

Grumman TBF/TBM Avenger (USA)

Appearing similar to the Helldiver, the Avenger prototype first flew in 1941 with its final delivery stretching into the mid-1950s. The model was also equipped with rear protection in the form of a manually-aimed .30-cal machine gun, along with a .30-cal in the upper forward fuselage and a single .50-cal gun in a dorsal power turret. The Navy model was a stalwart in the Pacific theater serving as a low-level torpedo attack aircraft.

Douglas SBD Dauntless (USA)

The armament of this dive bomber started in early versions with a single .50-cal Browning machine gun fixed in the nose. In typical Navy attack bomber style, there was also rear protection with one, and later two, .30-cal Brownings manually fired by the second crew member.

Grumman F4F/FM-1/2 Wildcat (USA)

One of the final prop fighters to carry guns in the fuselage, the Navy Wildcat had a pair of Colt-Browning .50-cal guns in that location along with four of the same in the expected wing location. Later versions would move all six guns to the wing leading edge location.

The super-slick P-51H was armed with six machine guns in the wing leading edge, all of which are visible in this photo. (Air Force Museum Photo)

P-59s had several different gun schemes, including a P-39 type cannon firing through the nose, along with other nose-mounted machine guns. (Air Force Museum Photo)

This P-61E Black Widow clearly shows its four 20-mm cannons solidly mounted in the nose. Some versions of this twin-engine fighter had a dorsal turret with four remotely controlled .50-cal machine guns. (Holder Collection)

Grumman F6F Hellcat (USA)

This stubby little fighter was one of the heroes of World War II with well over 12,000 delivered during the war years. The F6F was loaded for bear with the standard armament being six .50-cal Browning guns, three in each wing. The follow-on to the Wildcat also had an alternative weapons suite with a pair of 20-mm cannons and four .5-in Brownings, either load certainly able to get your attention.

Vought F4U Corsair (USA)

One of the real Allied heroes of World War II, the 400 mile per hour gull-winged fighter was a killer air defense fighter operating mostly off carriers. The technology of the model would allow it to also participate in the Korean conflict.

Initial versions carried an array of six .50-cal Browning machine guns.

Hawker Hurricane (British)

One of the two planes that helped win the Battle of Britain(the other being the Spitfire) the Hurricane first flew in 1935) with over 14,000 produced. A highly-flexible mission machine, the Hurricane performed in both air-to-air and air-to-surface missions.

During the early days of the war, the Hurricane was an excellent bomber destroyer with a murderous cone of fire from its eight .303 Brownings each with 333 rounds. But there were actually provisions for 12 guns.

For the ground attack mission, the plane was outfitted with a pair of 40-mm cannons, a role that it served with distinction against German tanks in North Africa.

The odd-looking P-75A Eagle never made it to production, but you better believe that it was armed like a bomber with ten .50-cal machine guns! (Vadnais Photo)

Here is a pair of the many .50-caliber machine guns carried by the research P-75A fighter. (Vadnais Photo)

The Navy SBD went back to an earlier era with this manually-operated .30-cal machine gun from an open rear-facing gunner position. (US Navy Photo)

The F4F Wildcat was the first of a tough family of Navy fighters. The gun armament varied with the model from two .50-cal machine guns in the fuselage, along with four and six machine gun wing configurations. (US Navy Photo)

The F4U Corsair was THE Navy fighter during World War II. The standard armament was six .50-cal Browning 53-2 machine guns (each with about 400 rounds) located in the unique folding wings. A variation occurred in the 1C version which sported four 20-mm cannons. (US Navy Photo)

Submarine Spitfire (British)

Along with the P-51 Mustang, the Spitfire was possibly the most famous fighter of WWII. Built in huge numbers, there was a massive number of versions to accomplish many missions. The Mark IB version had a pair of 20-mm Hispano cannons and four .303-in machine guns while the initial IA showed eight machine guns.

The Mark VA had eight machine .303-in machine guns, while the Mark VB and VI showed a pair of 20-mm cannons and four .303-in machine guns. In order to counter the potent German FW-190, the Mark IX was equipped with two .5-inch machine guns and a pair of 20-mm cannons. The Mark XIV version was built to go after the German V-1 buzz bombs, destroying over three hundred during the war. The armament was the same as the IX version, but the fuselage was considerably modified for the new mission.

Following the war, the Spitfire flew on with the Mark 21 toting four 20-mm cannons. Naval versions of the Spitfire, coined the Seafire, shouldered on and participated in the Korean War.

The F6F Hellcat was next in line with the standard armament being six .50-cal Browning machine guns mounted in the wing leading edges. (US Navy Photo)

Browning .303-inch machine guns were the guns for the Hurricane Mark I. (Air Force Museum Photo)

The four gun openings in the leading edge of the near wing of this Hurricane MK VII indicates its eight machine gun armament suite. (Air Force Museum Photo)

Loading of a Hurricane in this wartime situation is shown. The ammunition is loaded through the top of the wing. (Air Force Museum Photo)

Hawker Typhoon (British)

The Typhoon was approaching the ultimate prop-driven fighter aircraft powered with an awesome 2,180 HP Napier powerplant which provided a top speed of 412 miles per hour. The armament was equally as awesome with 12, count 'em-an even dozen, .303-in Brownings in the initial 1A version. Later versions would carry four 20-mm Hispano cannons in the outer wing areas.

Bristol Type 156 Beaufighter (British)

It looked somewhat stubby and awkward but the Beaufighter proved its worth on many occasions. With an attack mission, the twin engine fighter carried a brutal package of four 20-mm Hispano cannons placed on the underside of the forward fuselage. On early models, the rounds were hand loaded with 60-round drums, but that technique would later be replaced with a belt-feed system.

But there was more, a heck of a lot more. First, there was a Browning machine gun located on the upper fuselage which was aimed by the observer. Then, in the outer left wing's leading edge, there was a pair of fixed .303-in Brownings and four more in the right wing. How would you like to face this machine in a head-on attack?

Boulton Paul P.82 Defiant (France)

One of the most interesting European fighters from an armament point-of-view was this French fighter. First of all, there was no forward-firing fixed guns. Instead, the large dorsal turret, not unlike US Navy torpedo bombers, was in place. The installation was large, being taller than the cockpit, was hydraulically operated with four .303-in Browning machine guns, each with 600 rounds of ammunition.

The gunner was situated behind the pilot and had a control column which he could move either right or left for to rotate the dorsal turret, fore and aft for lowering and raising the guns, and a safety/firing button on the top.

Illyushin Il-2 Stormovik (Soviet Union)

First flying in 1939, the Il-2 and its follow-on versions played heavily on the Russian front throughout the war. Built strictly

A close-in view of the Spitfire with its 20-mm cannon clearly visible. (Air Force Museum Photo)

Note the high-visibility cockpit of this later-model Spitfire. The magnetic machine was armed to the teeth. (Air Force Museum)

A view of the characteristic 20-mm cannon, one of which that was carried on each wing of the Spitfire. Other armament included .50-cal machine guns. (Air Force Museum Photo)

The British Typhoon was considered by many to be the ultimate of prop-driven fighters. Its armament was awesome with a pair of 20-mm cannons on each wing. (Air Force Museum Photo)

as a close-support aircraft, the model was underpowered with only a 280 mile per hour top speed.

With a two-man crew version, the plane had rear protection in the form of a manually-aimed 12.7-mm BS machine gun in the rear cockpit. Up front armament consisted of two 20-mm ShVAK cannons and underwing racks for eight 82-mm rockets. A later version of the Il-2 would be more heavily armed, sometimes carrying four 20-mm cannons and two 7.62-mm machine guns in the rear dorsal rear turret.

Lavochkin LA-11 (Soviet Union)

The final Lavochkin fighter was by far its best, a fighter that would also later be used in the Korean conflict. The firepower came from right up front with four 20-mm ShVAK cannons that were arranged around the top of the front of the engine cowling with one version, and a later version going to three NS23, two on left, one on right. Interestingly, there were no guns carried on the wing leading edges. It was a technique that was also used with earlier Lavochkin fighter designs.

The British Blackburn "Roc" fighter carried an interesting bomber-style turret just aft of the cockpit. (Air Force Museum Photo)

The Spitfire Mark VB's armament consisted of two 20-mm and four .303-inch machine guns. (Air Force Museum Photo)

The Spitfire Mark IX version of the Spitfire was armed by two 20-mm and two .50-caliber machine guns. (Air Force Museum Photo)

The Me-109 Nazi fighter was one of the mainstays of the Luftwaffe and had formidable forward firepower in the form of three MG 17 7.62-mm machine guns. One of the gun's barrel came directly through the propeller nose cone, while the other pair was mounted on the upper fuselage directly in front of the cockpit wind screen. (Air Force Museum Photo)

The upper fuselage location of one of two 13-mm machine guns can be clearly seen in the photo of this wrecked FW-190. (Air Force Museum Photo)

Another weapon carried by the FW-190 was a pair of 20-mm cannons, one of which can be seen in the wing root location, as this example is being restored. (Air Force Museum Photo)

This captured Me-109 clearly shows the location of the upper fuselage machine guns. (Air Force Museum Photo)

Yakovlev Yak-3 (Soviet Union)

The Yakovlev designers brought forth the same firing-through-the-prop technique with their Yak-3 single engine fighter. The fighter would prove itself to be an effective aircraft capable of facing the best the Germans could mount.

It should be noted, however, that the center-mounted 20-mm ShVAK cannon was actually protruding out of the hub, and in front of the propeller. The twin 12.7-mm BS machine guns, though, were mounted on the upper fuselage, directly in front of the cockpit. Each of those latter guns carried only 250 rounds.

The Messerschmitt Bf-109 (Germany)

First flying in the mid-1930s, this fighter proved to be one of the dominant fighters in World War II. Through the course of the war, the model demonstrated many different weapons combinations. One of the most interesting gun mounting was

Although the Me-110 looked more like a light bomber, it was officially classified as a fighter. Its strong point was its forward-firing capability with two 20-mm cannons and four 7.92-mm MG 17 machine guns in the nose. (Air Force Museum Photo)

This Me-110 is shown in a maintenance hangar during WWII somewhere in Europe. (Air Force Museum Photo)

With its speed being the great adversary of the Me-163 Komet, firepower was minimal with only a pair of 30-mm MK 108 cannons, one in each wing root, as can be seen in this photo. (Air Force Museum Photo)

The twin machine guns of the rear turret installation on the well-armed Me-110. (Air Force Museum Photo)

the retaining of the firing-through-the-prop set-up of earlier days. The arrangement involved three 7.92-mm MG 17 machine guns, two located above the engine and firing through the prop and one through the propeller hub.

Later versions were incorporate potent 20-mm cannons in the wings. Some versions would have as many as three 20-mm cannons. Later versions of the Bf-109 would also incorporate rocket pods to augment the model's firepower. The model served as a huge threat to Eighth Air Force bombers from England as they attacked Europe. Some 35,000 were produced.

Focke-Wulf Fw 190 (Germany)

With its somewhat squatty appearance, the Fw 190 wasn't initially considered that much of a threat by the Allies. That would be a serious mistake, though, as the fighter challenged

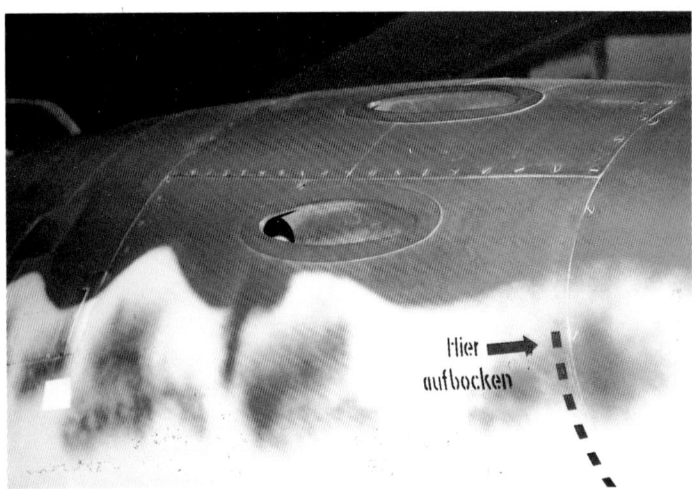

The second Nazi fighter of World War II, the Me-262, used the same cannon power as the smaller Me-163, only in this case, there were four of the 30-mm MK 108 cannons with two having 100 rounds and the other pair only 80. (Air Force Museum Photo)

Fortunately for the Allies, these damaged Me-262s were caught on the ground. The location of their nose-mounted cannons is clearly visible in this photo. (Air Force Museum Photo)

This particular Zero is carrying twin 20-mm cannons on each leading edge. (Air Force Museum Photo)

the Allies' best throughout the war. It carried an awesome weapons suite. The model first flew in 1939.

Armament was similar to the Bf-109 with a pair of 13-mm MG 131 machine guns located above the engine and firing through the propeller. Also, in the wings were a four 20-mm MG 151 cannons, two in the wing roots and two more further out on the wings. There were also provisions for a 30-mm Mk 108 cannon to be mounted to fire through the propeller hub.

Junkers Ju 87 Stuka (Germany)

This fixed landing gear attack fighter was best known for its howling engine scream as it made its vaunted ground attacks during the European conflict. And although the carrying of bombs was the main mission, there was also adequate armament for both front and rear protection. The guns on this machine were definitely for defensive purposes because the Stuka definitely wasn't any air-to-air weapon. Total production of this awkward looking plane was about 5,700.

Armament consisted of a pair of 7.92-mm Rheinmetaill MG 17 machine guns, one in each wing and an open MG 15 gun that was manually controlled from a closed position in the rear cockpit. In an interesting innovation, later versions of the Stuka carried modern-style weapons pods which could carry either multiple machine guns or 37-mm BK cannons.

Messerschmitt Bf 110 (German)

The versatile Bf 110 had several mission capabilities including day and night bombing missions along with ground attack at times. Its firepower, therefore, was directed in the forward direction with two 20-mm Oerlikon MG FF cannons and four Rheinmetall 7.92-mm MG 17 machine guns, all located in the nose area. Rear protection was provided by a 7.92-mm MG 15 manually aimed machine gun facing rearward out of the back of the cockpit bubble.

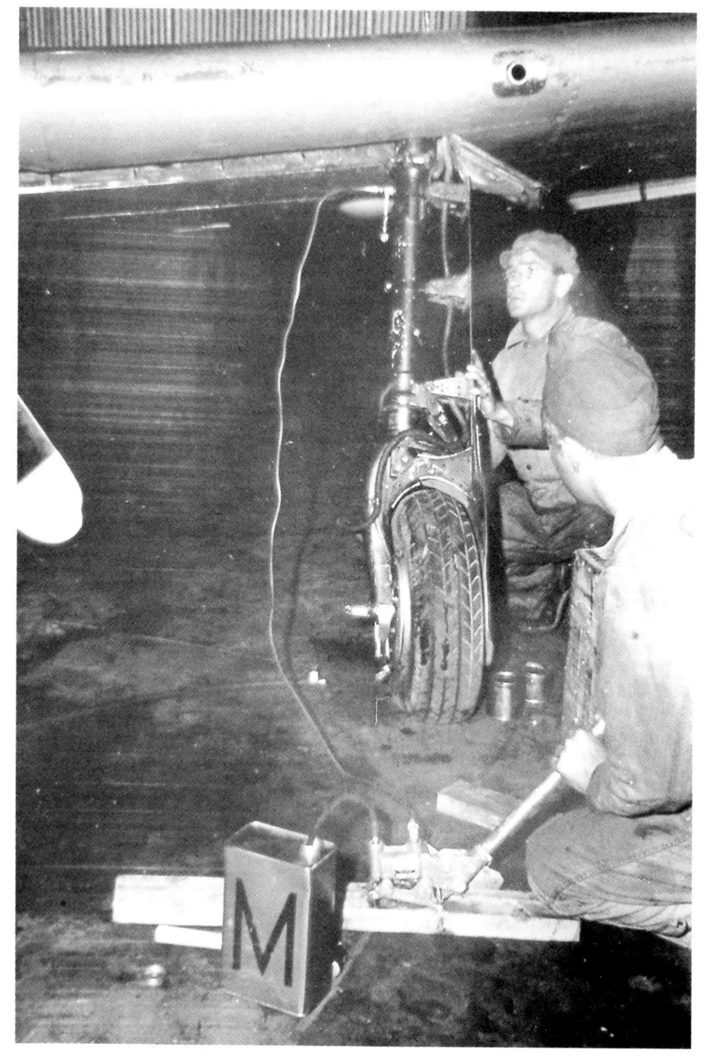

A view of one of the A6M Zero's 20-mm Type 99 cannons' locations, directly above the landing gear installation. (Air Force Museum Photo)

This Me-262 rests in a hangar near the end of WWII. Sporting obvious battle damage, this plane probably never flew again. (Air Force Museum Photo)

This A6M5 Zero was captured for evaluation and carries the USAAF markings. This later Zero version carried an optional fuselage-mounted 13.2-mm machine gun on the fuselage. (Air Force Museum Photo)

Messerschmitt Me I63 Komet (Germany)

Considered probably the most radically-designed fighter aircraft of World War II, the Komet was a rocket-powered short-duration fighter who's speed approached 600 miles per hour.

Since missions were short, usually just one pass being accomplished, there were only two guns, MK 108 cannons, located in the wing leading edge. Interestingly, each cannon only had 60 rounds on board. Their main targets were Allied bomber formations upon which they swept in, fired their rounds, and were out of range many times before a shot could be fired at them.

About 370 Komets are assessed to have seen service in their short career near the end of the war.

Messerschmitt Me 262 (Germany)

Military historians have indicated that had the Germans had the Me 262 earlier in the war, the victory would have come a lot tougher. Fortunately for the Allied effort, the few numbers of the plane didn't stop the momentum to victory, but the first operational jet-powered fighter played havoc with bomber formations. The approximately hundred of the model is assessed to have destroyed over 100 Allied bombers.

Its speed, of course, was its huge advantage but there was also great firepower available. With no propeller to synchronize through, the Me 262 mounted four 30-mm MK 108 cannons in the forward nose. Some versions also carried 50-mm rockets in addition to the cannon suite. In a unique offensive arrangement, certain versions also carried a dozen rifled mortars in the nose. With its array of offensive weapons, the Me 262 had a devastating stand-off capability against bomber formations.

Mitsubishi A6M Zero-Sen (Japanese)

The most-famous of the Japanese fighters was a single-seat carrier-based interceptor, known throughout the world as the Zero, a fighter that presented considerable problems for the US Navy in the Pacific theater.

Although being far from sleek and a speed far short of the P-51 Mustang and other Allied fighters, the carrier-based Zero made up for that deficiency with outstanding maneuverability and considerable firepower. Total production was almost 11,000.

The armament was a combination of machine guns and cannons. The pair of 20-mm Type 99 cannons was carried, one in each wing, with a 60-round drum in the outer part of the wing. In addition, there were also two Type 97 machine guns located above the front fuselage, each with 500 rounds. There would be several other variations in follow-on models.

This crashed Japanese Zero fighter shows a number of kill markings including a B-17 and P-38 among a number of others, possibly a B-25. If those claims are legitimate, the firepower of this A6M2 was extremely effective. (Air Force Museum Photo)

Boeing B-17 Flying Fortress (USA)

Ask many World War II historians which Allied aircraft contributed the most to winning the war, and for many, the answer will be a resounding response of the Boeing B-17 Flying Fortress.

This bomber evolved through the war years with the armament growing in numbers of guns and the sophistication of those weapon systems. The initial XB-17 carried only five .30-cal machine guns, a number that would increase greatly in the models to follow.

The armament would be increased for the C version, the breakout typically being seven machine guns, with one .300-cal and six .50-cal guns. The Flying Fortress was starting to

Although the B-15 would never make it to the combat skies of World War II, much of its armament technology would find its way into future bombers. (Air Force Museum Photo)

This Martin B-10 bomber was armed with a 3.30-cal machine gun in its large nose section. (Air Force Museum Photo)

The B-15's side protection came from a pair of side-fuselage mounted blister each carrying a single machine gun. (Air Force Museum Photo)

The clear turret mounted on the front of the B-10 bomber is clearly visible from this side photo. (Air Force Museum Photo)

It was a high mount for the top-fuselage gunner on the experimental B-15 bomber. It set the trend for armament on the bombers to follow. (Air Force Museum Photo)

The front end armament for the B-15 was interesting in the forward-firing machine gun was in a turret that was actually sitting on top of the larger front nose turret. (Air Force Museum Photo)

bristle with firepower covering all fields of fire where attacks would come.

The next major evolution came with the introduction of the B-17E variant which incorporated a redesigned rear fuselage along with the addition of mid-fuselage waist guns and three turrets. The variant supported eight machine gun locations. The turret locations included the forward fuselage just behind the cockpit, on the lower mid-fuselage, and in the new rear stinger location. That rear position had been woeful lacking in protection in earlier models. The E version was the first Flying Fortress produced in any significant numbers, 512 to be exact.

The B-17F continued to increase the firepower protection with this model now carrying nine .50-cal machine guns. The model featured the introduction of the advanced Bendix ball turret. The F version was produced by two other manufacturers besides Boeing, i.e. Douglas and Vega.

The front end firepower of the B-17E is shown in this photo under maintenance. The arrangement would be replaced with a power driven chin turret in the G model. (Holder Collection)

The use of upper turrets was becoming standard armament as can be seen on this XB-19 prototype bomber. (Air Force Museum Photo)

The upper turret on later model B-17s was a great protector of the Flying Fortress from above attacks and took out its share of enemy fighters through the war. The armament featured a pair of Browning .50-cal machine guns in a power-driven configuration. (Holder Photo)

They called it the chin turret and its pair of remotely-controlled .50-cal machine guns helped protect the front of the B-17G version, long a favorite enemy attack location. Combined with the firepower from the four guns of the upper and lower ball turrets, a tremendous firepower avalanche could be placed on forward approaching targets. There was also an additional optional gun in the normal nose position in addition to the chin .50-cal guns. It was the final version produced, and by far the most numerous with a total of 12,731 being produced by the three manufacturers.

But even with the significant firepower of the later versions of the B-17s, there was still deliberations about additional protection. To that end, a B-17-based B-40 'escort bomber' was designed with the sole purpose of providing firepower support to bombers on European missions. There was no standard armament for these planes which carried no

The lower ball turret provided the only lower protection for the B-17, shown here in a test firing. Twin .50-cal machine guns were again the gun array. (Air Force Museum Photo)

This famous Flying Fortress, the Shoo Shoo Baby, shows the awesome power-driven chin turret with a pair of .50-cal Browning machine guns. The chin turret was added because of front-end vulnerability to attack with earlier models. (Holder Photo)

Looking directly up to the chin turret of this B-17G, the details of the advanced power driven device can be seen. (Vadnais Photo)

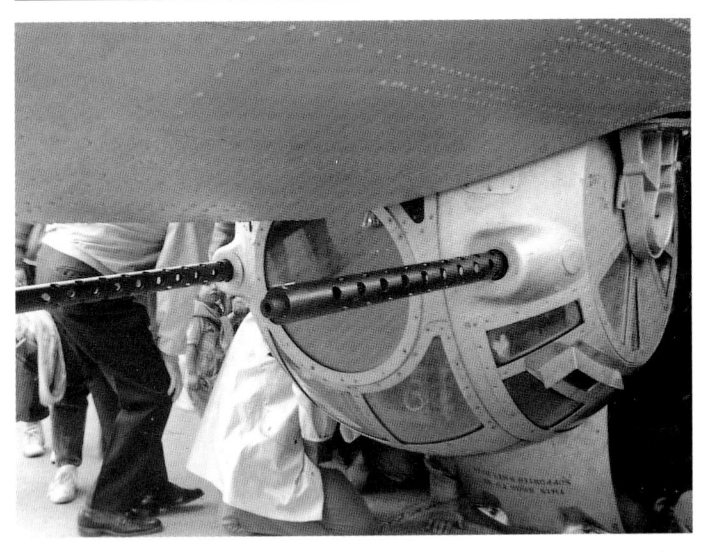

Without doubt, the B-17's lower ball turret was in a vulnerable position. The turret was heavily protected with armor plating. (Holder Photo)

The tail turret, also power driven, was enough to discourage bandits from attacking from the rear of the bomber. (Air Force Museum Photo)

bombs. Some versions carried as many as 30 machine guns and cannons. The machine guns were mounted in existing and additional ball turrets along with 20- and 40-mm cannons located in various positions.

The experiment did not prove to be successful since the B-40s were unable to keep up with the bombers on the way home after the mission aircraft had dropped their bombs. The YB-40s still had the considerable weight of their guns and ammunition to carry back to home base. Some of the techniques employed by the YB-40, though, would later be incorporated in the B-17F version, most importantly being the twin .50-cal chin turret installation.

Consolidated B-24 Liberator (USA)

Like the B-17, the firepower on the Liberator went from very primitive in the early versions, but by the end of the war, this famous bomber was armed to the hilt.

The final versions of the high-wing bomber carried ten .50-cal Browning machine guns in four electrically-operated turrets(Consolidated or Emerson in nose, the Martin Dorsal location, Briggs-Sperry retractable ventral ball and Consoli-

Tight formations were the keynote of protection for B-17 squadrons flying over Europe. It was imperative to bring every gun into play against an enemy attack, and it worked. (Holder Collection)

Early in the war, fighters (such at this P-51), didn't have the range to accompany bombers all the way to the target, leaving them to fend for themselves after the fighters had to turn back. Later, with wing-tip fuel tanks, the fighters could go all the way with the bombers providing attention to enemy fighters. (Holder Photo)

dated or Motor Products tail. The side positions were manually-operated positions on a single gun each.

The biggest improvement for the model was the addition of that front turret which corrected one of the plane's major deficiencies—front-on attacks from Nazi fighters.

Although the model had somewhat of a bad reputation with many pilots, being a lot more difficult to fly that the B-17, it was still built in greater numbers.

North American B-25 Mitchell (USA)

First flying in 1940, the B-25 proved to be a potent performer in a number of different missions. But one of the most important ground attack missions came from the machine gun and cannon complements the various models would carry through the war years.

The armament started off minimally with only two guns (a .30-and a .50-cal). It got better with the B version with twin .50-cals in an electrically-controlled dorsal turret and a ventral turret. Then everything went nuts when the G version was equipped with a 75-mm M-4 cannon which had to manually loaded. As with other Allied bombers, though, front end protection was also needed, more than the pair of .50-cal guns already in place. To that end, a package of four .50-cal guns were emplaced on the sides of the fuselage.

But it got even more awesome with the H version. In this variant, in addition to the M-4 cannon, an unbelievable pack-

It was an interesting gun experiment, the XB-40, which was in effect a 'B-17 Gun Ship', a Flying Fortress which carried a formidable array of guns for fleet protection. Its extra weight was the main reason for its failure. (Air Force Museum Photo)

Several companies produced versions of the B-40 Escort Bomber, this particular example being produced by Lockheed. (Air Force Museum Photo)

In addition to the chin turret, there were also a pair of hand-aimed .50-cal machine guns located on the sides of the forward fuselage. (Vadnais Photo)

The B-24's top turret provided protection above the Liberator. The power-driven turret mounted a pair of .50-caliber machine guns. (Air Force Museum Photo)

There was lots of experimentation during WWII. Here, a B-24 is equipped with a B-17 nose. (Air Force Museum Photo)

The last Consolidated Vultee B-24 Liberator is shown in final production. A number of the bomber's ten .50-cal Browning machine guns can be seen, here in the nose and upper fuselage turret. (Air Force Museum Photo)

This could have been the most famous and tragic B-24 of World War II. After crashing in the Libyan desert, the Lady Be Good was undiscovered for many years. Yet, its guns were still in firing condition. Here a member of the search crew looks at the .50-cal machine guns in the tail turret. (Holder Collection)

The tail stinger of the B-24, which sported a pair of .50-cal machine guns, looked like an afterthought being added to the design at the last minute. (Air Force Museum Photo)

This B-25G was a vicious as the shark teeth indicated. In addition to the normal armament, the nose also sported a 75-mm M-4 manually-loaded cannon with about 21 rounds on board. In addition, there were also a pair of .50-caliber machine guns for flak suppression. (Air Force Museum Photo)

age of 14 .50-cal machine guns formed probably the most awesome armament package of any World War II bomber. Eight of the guns fired in the forward direction. Needless to say, when one of these bombers made a ground attack, it definitely got the attention of the enemy, an ultimate hose-down. You'd definitely have to call the B-25G/H versions some of the earliest "Gunships."

Douglas B-26 Invader (USA)

For a light bomber, the armament for the B-26 carried an amazing amount of armament with ten .50-cal Browning machine guns, with an amazing number of six fixed in the nose! There were also two each in dorsal and ventral turrets.

There was another configuration(the B-26K) which carried eight nose guns or four nose-mounted cannons along with six .50-cal machine guns in the wings. That's 14 guns putting lead on the target at the same time. Definitely a situation that would get your attention.

Although much of the aviation history books will talk about the performance of the B-26 during World War II, 450 of the model (which would now be called the A-26) would perform in the Korean War.

Boeing B-29 Superfortress (USA)

Technology was moving fast on the armament front and the B-29 was one of the recipients. The marvelous new bomber

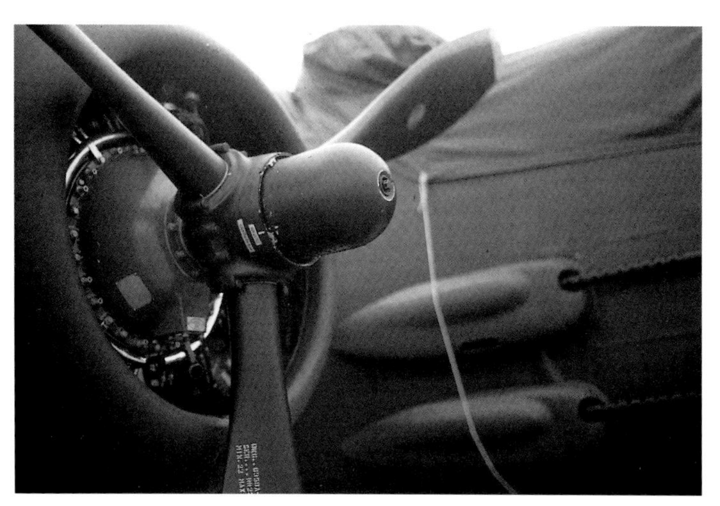

Certain versions of the B-25G carried so-called 'Package Guns', actually four .50-caliber machine guns, along the sides of the fuselage. Ground attack was definitely the name of the game with this fighting machine. (Holder Photo)

Normal gun array for the B-25 was a pair of .50-cal machine guns in the nose, twin machine guns in the tail stinger and the upper turret and manually-aimed guns in the side waist positions. (Air Force Museum Photo)

The top turret for the B-25 was similar to that of other bombers of the period with twin .50-cal machine guns. (Air Force Museum Photo)

The B-25 was the recipient of many different gun arrays. This field modification shows a B-25 carrying an amazing half dozen .50 cal machine guns. Bet the crew felt the vibration when they were all fired together! (Air Force Museum Photo)

was aerodynamically clean as was the armament it carried. No longer were there open gun positions on the fuselage sides and remotely controlled guns were now the standard.

Two GE turrets, one above and one below the fuselage mounted a total of four .50-cal machine guns. No gunners at the positions, though, with sighting and aiming accomplished from the nose or three waist sighting positions. Out back, there was awesome protection with the manned Bell tail turret mounting a single 20-mm cannon and twin .50-cal machine guns. Surprisingly, there was no front-facing armament.

Douglas A-20 Havoc (USA/France/Great Britain)

Often confused with the B-26 light bomber, the A-20 attack bomber was an entity on its own and performed well flying missions for three different countries.

A number of different versions were built of the A-20 with the G version being used extensively by US forces. That version carried four 20-mm cannons and two .50-cal machine guns, or six .50-cal MGs in the nose, dorsal turret, and ventral location. Formidable with either option, to be sure.

Earlier versions of the model had a number of different defensive suites. Most carried only the .303-in Browning machine guns with one aimed manually from the rear cockpit location (memories of World War I). The Boston III version, though, did carry four Hispano 20-mm cannons in a belly tray.

No caption for this slide or the slide labeled 2-115. This appears to be the front gun armament on a Junkers model. If this is too little to go on, captions will be placed in 1st proofs.

Forward-firing machine gun installation on the fuselage of a B-25. (Holder Photo)

The B-25B had a manually-aimed .30-caliber machine gun in the nose, an armament that would later be upgraded.

This manually-aimed .50-cal machine gun was one of a pair of waist positions on early model B-26 Marauders. (Air Force Museum Photo)

The tail armament for the B-26 was substantial with a manually-aimed .50-cal machine gun in the tail stinger. Also, the top turret, located much further back on the fuselage, sported a pair of .50-cal guns. (Air Force Museum Photo)

Different versions of the B-26 carried either one or two machine guns in the nose. This version carries the single .50-cal machine gun. (Holder Collection)

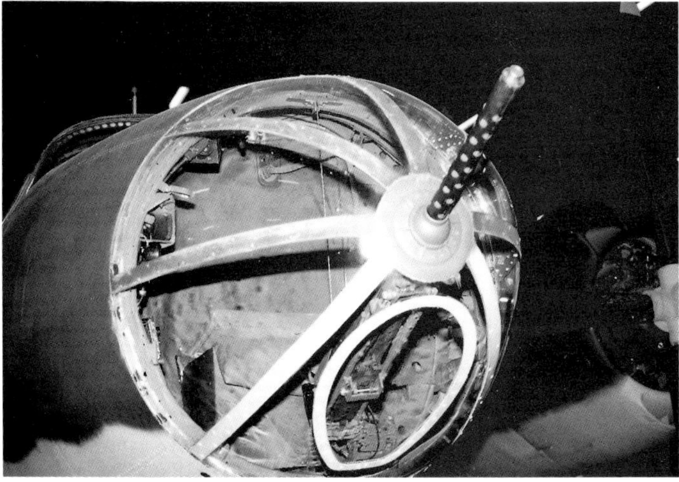

The single .50 caliber machine gun of this B-26B is clearly demonstrated in this photo. Note the swivel set-up in the center of the nose. (Vadnais Photo)

The gun array of this B-26B performed its mission as the plane is still flying even though there is significant battle damage clearly visible. (Holder Collection)

The upper ball turret of this B-26G contains a pair of .50-cal machine guns, two of up to 11 guns on various versions of the Marauder. (Vadnais Photo)

The B-26K 'Counter Invader' was a special model armed to the hilt for ground attack situations. Here, the eight .50-cal machine guns are shown on their nose mounting. It was firepower plus! (Vadnais Photo)

Douglas A-26 Invader (USA)

In an era when aeronautical technology was moving at a frantic pace, this particular model saw action in three wars (World War II, Korea, and Vietnam.) The model would later be named the B-26, but during World War II, it carried the 'attack' designator.

One of the early prototypes carried a massive 75-mm gun and another with four 20-mm forward-firing cannons and four .50-cal guns in the upper turret. (See B-26 discussion for further evolution of the A-20 model).

Curtis SB2C Helldiver (USA)

Even though the Helldiver might look like a fighter, it was officially classified as a 'dive bomber' and will hence be carried in the bomber category here.

Forward firepower was the highlight here with either a pair of 20-mm or four .5-in guns located in the wings for a significant punch. There was also some rear protection with either two .30-cal or a single .50-cal machine gun in the rear cockpit. Over 7,000 of the carrier-based dive bomber were produced.

Avro 683 Lancaster (British)

This seven-seat heavy bomber was one of the stand-by's for the Allied effort during the war. A four-engine configuration, the Lancaster had a terrific bomb-load capability and was armed to the hilt.

The model featured three gun-mounting turrets with the nose unit mounting a pair of Browning .303-in MGs, the dorsal turret with the same array, while the tail turret doubled that complement with an awesome four Brownings. In some versions, the dorsal location (a Martin turret) mounted a pair of .50-cal machine guns. The only lacking of protection was from underneath the bomber where there were no guns.

Totally unexpected wing guns were also a part of the B-26K's armament package. (Air Force Museum Photo)

The four .50-caliber machine guns of the forward upper turret of this B-29J are clearly visible in this photograph. (Air Force Museum Photo)

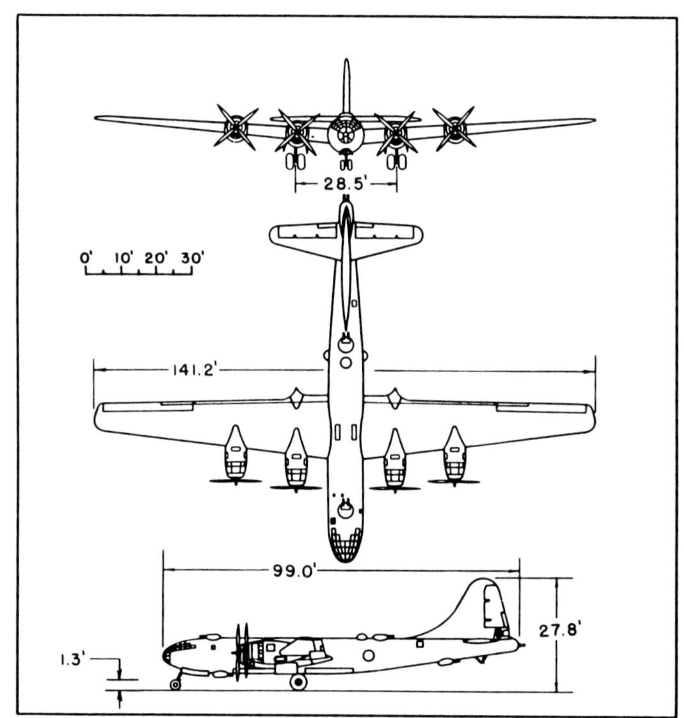

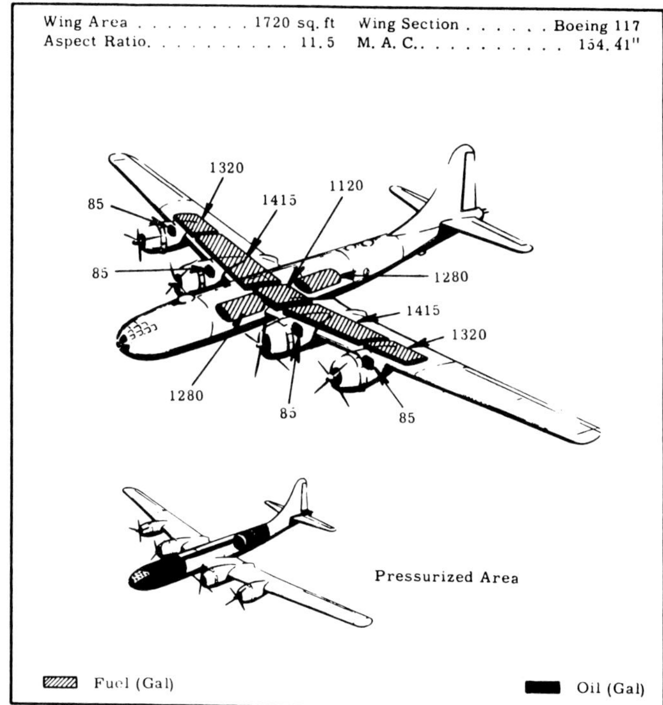

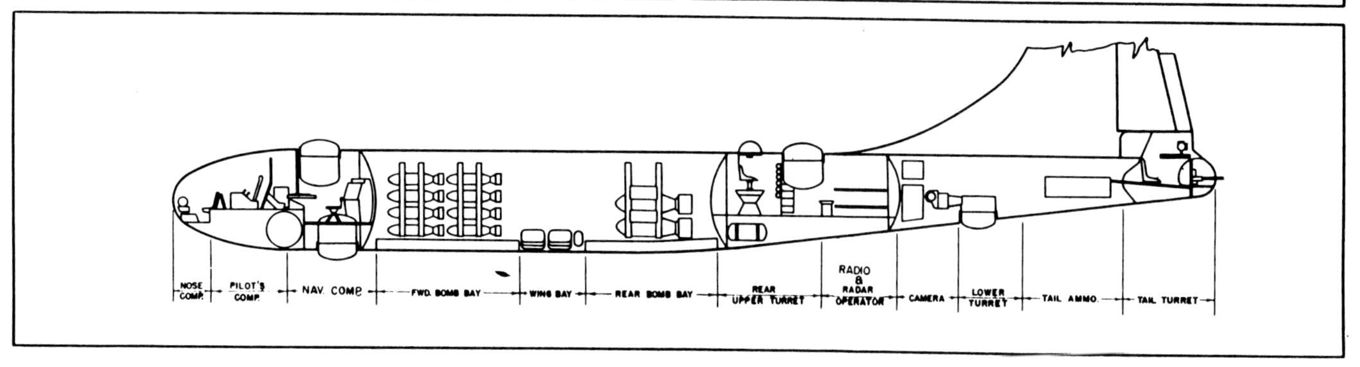

Schematics for the B-29A model Superfortress. (USAF drawing)

The B-29 was a heavy player during the Korean War. The Command Decision was one of those Superfortresses, and from the five MiG symbols under its nose, you can see that this B-29 was indeed an ace. (Holder Collection)

A A-20 Havoc fires its machine guns on a gunnery range at Wright Field, in August 1944. (USAF Photo)

Handley Page Halifax (British)

It wasn't as famous as its famous running mate, the Lancaster, but the Halifax proved a worthy participant in the big war. The four-engine bomber carried a crew of seven and had ample firepower to battle the hordes of German fighters it would face.

There were a number of different versions with the last being delivered in 1946. There were usually a pair of Browning .303-in machine guns in the nose, four in the tail turret, and a pair in the manual beam positions.

de Havilland Mosquito (British)

The legendary wooden construction Mosquito was one of the real heroes of World War II with many versions produced. It was a bomb-toter—true—but its real claim to fame was its forward firing front firepower.

It was all mounted on the forward portion of the fuselage with four .303-in machines(with 2000 rounds each) protruding from the nose and four 20-mm Hispano cannons(with 300 rounds each) located just below and slightly aft on the lower fuselage.

The fuselage-mounted machine guns on the side of the fuselage of this A-20 can be seen protruding from the lighter-colored protrusion. (Air Force Museum Photo)

The firepower of the A-20G is vividly illustrated by the .50-cal machine gun array shown here. (Vadnais Photo)

Two of the A-20G's .50-cal machine gun array were contained in this power-driven upper turret. (Vadnais Photo)

The firepower of the A-26A is shown in this war years firepower test. (Air Force Museum Photo)

The A-26 Invader was capable of carrying up to eight .50-cal machine guns. Here is the location of two of them in a pod mounted underneath the wing. (Vadnais Photo)

Vickers-Armstrong Wellington (British)

This outstanding WWII bomber was a favored British weapon with well over 11,000 produced.

The armament for the two-engine light bomber was significant but it was only for the front and rear of the aircraft. The plane carried nearly-identical Vickers turrets with each carrying a pair of Vickers .303-in machine guns.

The Mark IA version would feature Nash and Thompson power turrets while the Mark IC version would add manually-operated side fuselage machine guns.

Heinkel He 111 (German)

One of the most highly-produced German bombers, the twin engine He 111 was a formidable Axis weapon during World War II.

One thing that could be said about the aircraft was that it was armed to the hilt in just about every direction.

First, there were usually a pair of MG 15 7.92-mm machine guns in the nose. The lower gun fired directly forward from the front of the nose section. A second machine gun was carried higher on the nose, a weapon that was installed in the field for added front protection.

On the top of the fuselage, a top gunner protected the plane from straight-down enemy attacks. In addition, there were also machine gun locations in both sides of the fuselage and a remotely controlled unit in the tail. But there was more. A interesting twin-gun protrusion was mounted below the fuselage just aft of the wing root. A machine gun fired rearward in addition to a forward-firing 20-mm MG FF cannon for anti-shipping operations.

Junkers Ju 88 (Germany)

One of the most widely-used German bombers of the war, the Ju 88 employed a most interesting gun installation.

The new name of the gun game with this light two-engine bomber was a so-called ventral addition located directly under the cockpit on the lower fuselage. Within this protrusion was an awesome package of four Mauser MG 151 machine guns with two hundred rounds each. Rear protection was provided by a machine gun in the ventral.

There was also a rear-firing machine gun in the aft portion of the high-sitting cockpit. Two additional machine guns fired forward, one from the lower nose location and the other from the front of the cockpit.

This A-26B carries possibly the ultimate ground attack weapon with a 75-mm cannon mounted on the right side of its nose. (USAF Photo)

The Mitsubishi G4M Betty was a Japanese land-based torpedo bomber produced in large numbers. Well armed, it carried three manually-aimed machine guns in the nose, dorsal and ventral positions and a 20-mm gun in the tail. (Air Force Museum Photo)

Chapter 3: Korea

The Korean War, though hot on the heels of World War II, saw a quantum leap in air warfare technology. Among the many advances, probably the most memorable was the introduction of jet fighters and fuselage-mounted guns.

The early years of the war, for both the United Nations forces and the communist forces, were marked by the use of WWII-era aircraft, some with heavy modifications, and the later introduction of the jet fighter. The North Korean Air Force was made up of front line WWII Soviet aircraft, including the Il-10, the Yakovlev Yak-3 and Yak-7 Fighters

The most famous of the Korean era fighters has to be the North American F-86 Sabre and the MiG-15. The Sabre, the Air Force's first supersonic fighter, was the class of the war. More than 5,000 were eventually made, and somewhere in the world, more than 200 continue to fly in the hands of a military or private owners.

The MiG-15's were flown by North Korean, Chinese and Soviet pilots during various portions of the war, and in the early years, the MiG-15 was measuring stick that all other fighters were measured against. After the introduction of the F-86, the Soviets tried to regain the technology edge with the MiG-15bis with a more powerful engine and better guns. Despite these upgrades, the F-86 still held the edge. For whatever reason, the Soviet bosses in Moscow refused to send the more advanced and lethal MiG-17 to Korean until the last weeks of the war and even then it was too little to late.

Two F-84s from the Ohio Air National Guard roll towards their target. (USAF photo)

An F-86H of the Maryland Air National Guard awaits an ammunition loader. This aircraft was assigned to the 104th Tactical Fighter Squadron. (USAF photo)

The early days of the war, where much was in doubt, the air defense of the peninsula belonged to Lockheed's F-80, North American's F-82 Twin Mustang and Douglas's B-26/A-26 bomber/attack plane.

They were up to the challenge during the summer of 1950. Performing close air support and ground interdiction missions behind enemy lines they virtually destroyed the North Korean air force of World War II vintage aircraft. But the arrival of the Chinese and Russian air forces in November 1950 changed things dramatically.

A photo that pretty much covers the era—a MiG-15 in the foreground, a F-86 behind it and way in the back is the nose of the B-36 Peacekeeper bomber. These three aircraft embody the Korean War era. (USAF Photo)

The North Koreans roared down the peninsula during the early weeks of the war, only to be pushed back to the Yalu by October of 1950. Then the Chinese Army jumped into the fray and pushed U.N. troops south of the 38th parallel by January 1951. The battle lines shifted, but by much smaller margins for the next two years until the cease-fire was signed. This latter part of the war resembled the trench warfare of WW I. Excursions by both sides, but no massive push to crush opposition. In this light, the air war accounted for many of the successes for the U.N. troops.

The aircraft of the air war were varied and ranged, as usual, from old yet effective to new and untested. The bulk of the air war was fought by U.S. forces. The major players in war follows.

The F-82 Twin Mustang waiting for a mission. The Twin Mustang was not two Mustangs fused together as the name would imply, but was new design that incorporated much of the design of the Mustang. (USAF Photo)

Chance Vought F-4U Corsair (USA)
Built primarily for Navy/Marine operations in the late 1930s, the F4U had a pretty rocky beginning. During carrier evaluations, the pilots complained that the long nose gave poor take-off and landing visiblity, and its engine generated a huge amount of torque at low speeds. An unwary pilot might shove the throttle forward, pull back on the stick. The aircraft, with its torque, would flip onto its back and dive into the ground.

While an outstanding fighter in WWII, the Corsair was unsuited for the jet age. Early during the war, the Corsair was used for close air support, train hunting and the like. As with most American fighter aircraft of this era, it is armed with six .50-cal machine guns and LOTS of ammo!

The Navy's F-8F Bearcat in traditional Navy paint schemes. (Vadnais Photo)

Grumman F-8F Bearcat (USA)
The Bearcat was the last in a long line of single seat, single engined fighters derived from the F-4F Wildcat. It had a very successful career in the Navy, flying in WW II and Korea. It was armed with four 4 .50-cal machine guns. Some later variations included a cannon.

Grumman F-9F Cougar (USA)
Grumman's first swept wing fighter, it was developed in 1951 as a carrier based fighter. It was armed with four 20-mm cannons. Nearly 2,000 Cougars were built and they saw action in Korea and in Vietnam. The Vietnam missions were mostly reconnaissance versions.

A F-80B over the skies of North Korean. (USAF Photo)

Grumman F-9F Panther (USA)
Yes, it's the same designation, but a VERY different aircraft. The Panther was a straight winged fighter that holds several notable achievements in Navy history. The Panther was first Navy jet to see combat, on July 3, 1950, in Korea. It had its first taste of blood on November 9, 1950 when a Panther shot down a North Korean MiG-15.

The Panther was armed with four 20-mm cannons in the nose, a number of 5-in rockets, and up to 500 pounds of bombs.

North American F-51D Mustang (USA)
Originally designed for the British in 1940, and excelled in every theater during World War II. By the time the Korean War came around, the aircraft had been re-designated the F-51. The Mustang was used primarily for close air support for ground combat troops until they were withdrawn from combat in 1953, another victim of the jet age. North American eventually built 15,261 Mustangs. During the Korean War, the Mustang was still armed with its original six .50-cal machine guns, and either 2,000 lbs of bombs or 10 5-in High-Velocity Aerial Rockets.

F-80B Shooting Star showing off three of its six .50-cal machine guns. (Vadnais Photo)

Another angle of the F-82 Twin Mustang on the ground in Korea (USAF Photo)

Lockheed F-80 Shooting Star (USA)
The 'Star was the Air Force aircraft to exceed 500 mph in level flight, the first jet to be used in combat, and the first mass produced jet fighter in the world. Designed my the famous Clarence "Kelly" Johnson (of U-2 and SR-71 fame), the Lockheed design was built around the British developed de Haviland H-1 turbojet. Designed in a week, and built in less than five months, the Shooting Star eventually became a pilots dream come true — quick, maneuverable, and forgiving. Designed during the second half of the second World War, the then P-80 never saw combat due to parts shortages and engine problems. It entered active duty in 1946 and the basic fighter (F-80A) never saw action, reconnaissance versions and F-80B and C models were used extensively in Korea. During the early months of the war, more than 15,000 sorties were flown. The worlds first all jet air battle took place on November 8, 1950, when 1st Lt. Russell J. Brown shot down a MiG-15. The F-80 soon was unable to fight on its own, and required extensive top cover from the F-86s as the war progressed. Records show that only 14 F-80s were shot down by enemy aircraft, but more than 100 were downed by ground fire.

The F-80 was eventually powered by an Allison J33 (5,400 lbs of thrust) and was armed with six .50-cal machine guns and 16 5-inch rockets or one ton of bombs. It had a top speed of 580 mph and a range of nearly 1,400 miles.

North American F-82 Twin Mustang
The Twin Mustang was an attempt to make the premier WWII fighter into something it couldn't be – a jet killer. The F-82's roots began in 1944 when the Army Air Corps began development of a long range bomber escort with room for two pilots, one to fly, the other for relief. Although the plane looks like two P-51 fuselages jammed together, it was really a clean paper design. The Twin Mustang, along with the venerable A-1 Skyraider, were the last propeller powered fighter acquired by the Air Force. Coincidentally, they are also the last of the tail draggers.

During the Korean War, the Twin Mustang were among the first USAF aircraft to see combat. Based in Japan, they flew nearly 2,000 sorties over Korea before being withdrawn from combat. During their brief combat career, the Twin Mustang had three aerial kills.

The F-82 was powered by twin counter-rotating Allison V-1710 engines. Armed with six .50-cal machine guns and 4,000 lbs of bombs or 25 5-in rockets, the Twin Mustang could reach 470 mph.

A KB-29 refuels two F-80 Shooting Stars during the Korean War. The Korean War was the first time aerial refueling was used during combat, and it was a godsend. It not only extended the combat range of aircraft, but many a pilot would have gone bingo fuel and punched out if not for the tanker. (USAF Photo)

The business end of the F-84 Thunderjet, with four of its six .50-cal machine guns showing. (Vadnais Photo)

Republic F-84 Thunderjet (USA)
The Thunderjet was the first jet fighter to begin its design after the conclusion of World War II, and began flying in 1947. It was the first American fighter capable of carrying a tactical nuclear weapon. Also, it was the last of the straight wing designed fighters, all subsequent designs being swept wing. The Thunderjets entered the Korean War in 1950 as B-29 bomber escorts. Later, they were used for ground attack. F-84 pilots flew more than 86,500 missions dropping more than 50,000 tons of bombs. The F-84 also pioneered the use of aerial refueling through extensive testing with KC-97s. Also, when the Air Force Aerial Demonstration Team, The Thunderbirds, formed, the first aircraft they used was the F-84 Thunderjet.

The F-84 carried six .50-cal machine guns and up to 6,000 lbs of bombs or rockets. It was nuclear capable, and could reach 685 mph with a range of 1,650 miles.

North American F-86 Sabre (USA)
Carrying six .50-caliber (or 12.7-mm in today's parlance) machine guns with 400 rounds of ammunition for each, in the nose, the F-86 could also handle up to four weapons pylons under the wings (F-86F). Later models replaced the machine guns with four 20-mm cannons and 600 rounds of ammunition.

The Sabre was the first of the swept wing fighters to enter Air Force service. It also became the first U.S. built fighter to break the speed of sound. During the Korean War, Sabre pilots downed 792 MiG-15s while losing only 76 – a 10-to-1 kill ratio. The last aerial kill of the war went to a Sabre pilot, Capt Ralph S. Parr, who downed a Il-2 on July 27, 1953. All 36 aces of the Korean War were Sabre pilots

Northrop F-89 Scorpion (USA)
An all weather fighter-interceptor, the Scorpion was the first fighter to lack a gun. It was armed with 52 Mighty Mouse 2.75-in folding-fin aerial rockets carried in two pods on the wing tips. The woefully underpowered Scorpion had a pretty benign career except for one career highlight. In an program called Operation Plumb Bob, an F-84J fired an AIR-2A Genie over Yucca Flats, Nev. This is the only time in history that air aircraft launched and detonated an air-to-air nuclear rocket.

Lockheed F-94 Starfire (USA)
Also armed with only rockets, the Starfire replaced the F-82s in Korea beginning in 1952. Jammed full of advanced electronics and fire control radar's, the Starfire was a combat aircraft forbidden to fly over North Korea, lest one get shot down and compromise the technology. Eventually, with mounting B-29 losses, this restriction was lifted. For the first time an aircraft downed another without ever seeing it when, on Jan 30, 1953 a Starfire fired on and downed a LA-9 fighter.

Mikoyan-Gurevich MiG-15 (Soviet Union)
The mainstay of the North Korean/Chinese Air Forces. The MiG-15 was a great little airplane—quick, maneuverable and with adequate guns. It's biggest downfall, in air-to-air combat with the F-86, was that it was slower in the transitions between the different air combat maneuvers. With each turn or twist, the F-86 would gain a slight advantage until it had enough for the kill.

The MiG-15 carried two 37-mm cannons under the nose.

Mikoyan-Gurevich MiG-17 (Soviet Union)
A late comer to the air war in Korea, the MiG-17 was a definite improvement over the MiG-15. However, the Russians hesitated so long in using it, that the war was all but over when it took to the skies.

The -17 carried three 23-mm NR-23 cannons and could carry up to a ton of bombs or rocket launchers.

Yakolev Yak-3 (USSR)
Designed and built during WWII, the Yak-3 was one of the fastest fighters of the war. As with many aircraft, it was quickly outclassed by the jet fighters during the war. It was originally armed with a 20-mm cannon firing through the propeller shaft. That was replaced with a 37-mm cannon later. It also carried two 12.7-mm BS machine guns on either side of the engine. The wooden wing was eventually replaced with a metal one by the time it saw action in Korea. More than 4,800 of these aircraft were built from 1943-1946.

A weapons load for the F-86 Sabre (USAF Photo)

Martin A-26/B-26 Invader (USA)

Originally designed as an attack aircraft the A-26 was re-designated the B-26 in 1948. When the Korean War broke out, it was the only attack aircraft available to the U.S. forces. The Invader flew its first combat mission on June 29, 1950 against targets in Pyongyang. The A-26/B-26 is credited with destroying 38,500 vehicles, 3,700 rail cars, 406 locomotives and seven aircraft on the ground. On July 27, 1953, the last air action of the war was accomplished by an Invader 24 minutes before the cease-fire was signed.

A Flying pillbox, the Invader was armed with 10 to 12 .50-cal machine guns (four or six in the nose, two each in the top and belly turrets, and 6,000 lbs of bombs internally and externally. The wings also had hardpoints were an additional four .50-cal machine gun pods could be mounted.

The Invader was so successful that it continued this ground attack mission into the Vietnam War

The F-86 Sabre could carry either four M-39 20-mm cannons, shown here, or six .50-cal. machine guns in the nose. (Vadnais Photo)

Boeing B-29 Superfortress (USA)

The penultimate bomber of the second World War, the B-29 saw action in the Far East, where its attributes long range and heavy bomb load could be best used. The Superfortress introduced pressurized cabins and remotely controlled gun turrets, a feature that shows up in later Boeing bombers. They are most famous for being the platform that dropped the two nuclear bombs on Japan. After the war, many were modified to aerial tankers. But when the Korean War broke out, many saw combat once again, being the only heavy bomber in the theater.

With a crew of ten, the Superfortress was armed with eight .50-cal machine guns in four remotely controlled turrets. Many were field modified to carry an additional two .50-cal machine guns and one 20-mm cannon, and up to three more machine guns in the tail. Its bomb load maxed out at 20,000 lbs.

Consolidated-Vultee B-36 Peacemaker (USA)

Originally designed to reach targets in Nazi Germany from bases in the United States should Britain fall to the Germans. It was the largest bomber in the world and could carry 72,000 lbs of nuclear or conventional weapons. It did not see action in Korea, but it stood alert duties in the U.S., deterring the Soviets from possibly getting more involved in the peninsula's conflict.

The Peacemaker came with 16 M24A1 20-mm cannons in eight remotely controlled locations – the nose, tail, and retractable top and belly turrets. Among the 16 crew members were five gunners in two locations (one in the nose, one in the tail) to control the guns.

North American B-45 Tornado (USA)

The Tornado did not see action in Korea, like the B-36. But, like the Peacemaker, it stood alert duty in Europe, doing its duty in keeping the Soviets' eyes focused on Europe, and not the Far East. It was the Air Force's first four-engine jet bomber to be produced. Three RB-45s saw limited duty in Korea as reconnaissance aircraft during their operational evaluation phase, but their combat time was extremely limited.

The Tornado was armed with two Browning M70 .50-cal machine guns in a tail stinger turret. The crew of four included the gunner.

The Aces

It is very difficult to account for the validity of the air-to-air kills claimed by the North Korean/Chinese/Soviet pilots. The numbers that they claim add up to nearly 10 times as many aircraft as the UN forces lost. Soviet claims alone are wildly exaggerated, claiming to have shot down 1,200 U.S. Air Force planes while the Air Force can account for only 139 air-to-air losses — 121 of them fighters, 18 bombers.

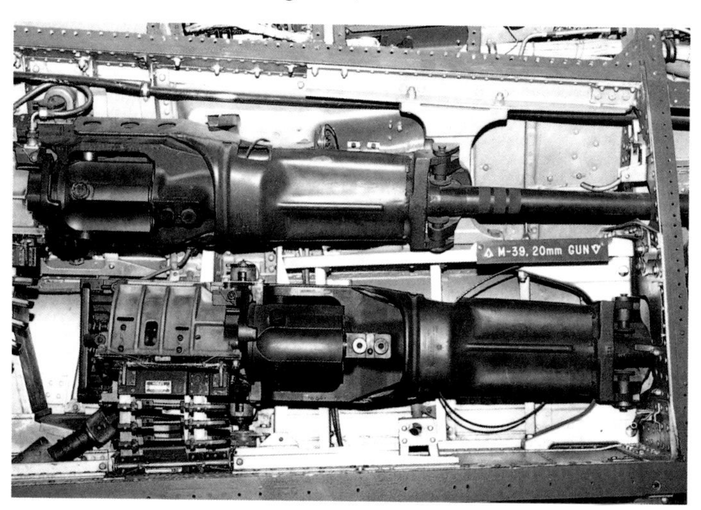

A close up to two M-39 20-mm cannons in the F-86. (Vadnais Photo)

An F-89 Scorpion on the tarmac. (USAF Photo)

Sabre pilots alone accounted for 792 MiG-15 kills and a total of 810 enemy aircraft shot down.

USAF/USN Aces of the Korean War

McConnell, Capt Joseph Jr.	16
Jabara, Maj James	15 (+3.5 in WW II)
Fernandez, Capt Manuel J.	14.5
Davis, Maj George Jr	14 (+7 in WW II)
Baker, Col Royal	13 (+3.5 in WW II)
Blesse, Maj Frederick	10
Fischer, 1st Lt Harold	10
Garrison, Lt Col Vermont	10 (+7.33 in WW II)
Johnson, Col James	10 (+1 in WW II)
Moore, Capt Lonnie	10
Parr, Capt Ralph Jr	10
Foster, Capt Cecil	9
Low, 1st Lt James	9
Hagerstrom, Maj James	8.5 (+6 in WW II)
Risner, Capt Robinson	8
Ruddell Lt Col George	8
Buttlemann, 1st Lt Henry	7
Jolley, Capt Clifford	7
Lilley, Capt Leonard	7
Adams, Maj Donald	6.5
Gabreski, Col Fancis	6.5 (+28 in WW II)
Jones, Lt Col George	6.5
Marshal, Maj Winton	6.5
Kasler, 1st Lt James	6
Love, Capt Robert	6
Bolt, Lt. Col. John F. (USMC)	6 (+6 in WW II)
Whisner, Maj William Jr	5.5 (+15.5 in WW II)
Baldwin, Col Robert	5
Becker, Capt Richard	5
Bettinger, Maj Stephen	5 (+1 in WW II)
Bordelon, Lt Guy P. (USN)	5
Creighton, Maj Richard	5 (+2 in WW II)
Curtin, Capt Clyde	5
Gibson, Capt Ralph	5
Kincheloe, Capt Iven Jr	5
Latshaw, Capt Robert	5
Moore, Capt Robert	5
Overton, Capt Dolphin III	5
Thyng, Col Harrison	5 (+5 in WW II)
Westcott, Maj William	5

Interesting Statistics:

Category	Air Force	Navy	Marine
Total missions flown	720,980	167,552	107,303
Bomb dropped (tons)	386,037	120,000	82,000
Aircraft losses (combat and other)	139	814	368
Enemy destroyed in air	900	16	35
Enemy destroyed on ground	25	36	n/a

With its eight .50-cal machine guns, the A-26 Invader was a fearsome ground attack aircraft. (USAF photo)

Chapter 4: Vietnam

With the huge technological edge the U.S. had during this war, you would think that even with the meddling by the politicians, we still would have won the war. But it was an object lesson to many of today's military leaders that if you're going to send a sledge hammer to war, use it as a sledge hammer, not a scalpel. Trying to bomb bicycle trails with huge B-52's now seems absurd, but it didn't to the political leaders of the time. And when they wanted bike trails bombed, all the Air Force had in its arsenal was sledge hammers. Today, with the advent of numerous fighter-bombers, precision guided munitions, it is possible to bomb a bicycle on a bike trail. But not then, and that speaks volumes as to the lack of success in Vietnam.

Airpower took the front seat in Vietnam, from close air support via the gunships, to air rescue operations, from SAM hunting to bridge dropping, air power was the most potent force in theater – when used properly. During Operation Linebacker and Linebacker II, the sheer destruction brought on by air power forced the Vietnamese to the peace talks, while the daily air tasking orders from Washington, D.C. showed the folly of trying run a war from the office, not the front lines.

As far as air combat went, it did not compare to previous wars. Only five men made it to "Ace" status – five kills. As a matter of fact, the USAF only had 137 official kills (45 with the gun as the primary weapon), the Navy about another 100. As was the tradition then, when a two-seat aircraft had a kill,

A-1E Skyraider, nicknamed "Sandy," at sunset over the Ca Mau peninsula in Vietnam. The Sandy was famous for its extremely large weapons load and long loiter time. (USAF Photo)

An F-100 Super Sabre rolls towards its target. Note the gun smoke stains on the bottom of the fuselage just behind the nose. (USAF Photo)

A KC-130 Hercules tanker, from the Marine Corps, refuels two Marine A-4 Skyhawks, nicknamed "Scooters." (USAF Photo)

F-4D Phantom, tail number 463 with six kills. Since aircrew did not often fly the same aircraft on every mission, this plane had several pilots/radar officers claim kills. The fifth kill for this jet was also the fifth by Capt. Steve Ritchie, making him the first American ace of the Vietnam War. It was his backseater, Capt. Charles DeBellevue's fourth kill. DeBellevue went on to score two more kills, making him the top ace of the war. (USAF Photo)

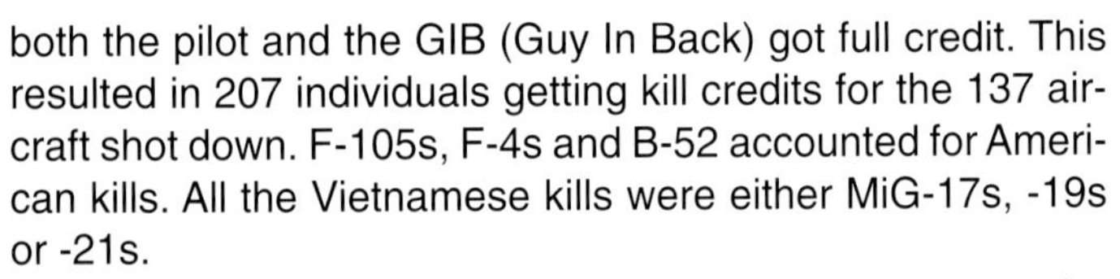

both the pilot and the GIB (Guy In Back) got full credit. This resulted in 207 individuals getting kill credits for the 137 aircraft shot down. F-105s, F-4s and B-52 accounted for American kills. All the Vietnamese kills were either MiG-17s, -19s or -21s.

Due to the politically motivated restrictions put on the warfighters by Washington, there were no aerial victories from Feb. 14, 1968 to Feb. 21, 1972, a rather long dry spell!

The war saw thousands upon thousands of American youth face an enemy that the vast majority didn't understand. They were following orders. In the air war, these young men were called upon to fight an enemy who, while seen as backward and deficient in technology, could still kill with the greatest of ease. And had the weapons (read surface-to-air missiles) to do it!

More than six million tons of munitions were used during the air war, with 2,257 aircraft lost to combat and other operational causes, costing about $3 billion.

Of the many types of aircraft used in the war, some for their originally intended purposes, others field modified to meet a need, the major ones are listed below.

An AC-47 on the ramp in Vietnam. Spooky rarely flew during daylight hours, preferring to do their work under cover of darkness. (USAF Photo)

Douglas AC-47 (USA)
Called the Gooney Bird, Puff the Magic Dragon, or simply Puff, the venerable C-47 was equipped with three M-61 .50-cal machine guns. These were later replaced by three side firing, electrically operated GAU-2/A Gatling-type six-barreled miniguns. The crew had to drop flares by hand to see the enemy at night

AC-119 Shadow (USA)
Shadow added a fourth gun and had a longer loiter time. this gunship had the four GAU-2B/A 7.62mm Gatling guns, and added two 20mm cannons and a searchlight to the deadly mix. It was also the first gunship to use onboard sensors to assist in finding and targeting the enemy. The biggest difference however, was that the K model had an additional two jet engines strapped on the wings that resulted in a 25 percent increase in its gross take off weight — more bullets, more flares, more fuel.

Lockheed AC-130 Spectre
The most advanced of the Vietnam era gunships, and still the basic platform used for today's gunships, the Spectre went into battle in 1967. Within three years, the single squadron of AC-130s at Ubon, Thailand had destroyed more than 5,000 heavy trucks carried men and material to the front.

(For more on the gunship, both fixed and rotary wing, see Chapter 6. The development of the gunship is dealt with in more detail there.)

Douglas A-1 Skyraider (USA)
This backbone of Navy aviation began its career in 1944 as a carrier based, dive bomber. It reflected the lessons learned in the Pacific war, namely carry a huge bomb load! By the time the production lines closed in 1957, 3,180 Skyraiders were built.

The A-1 has to be the most versatile fighter ever built. With different quick change kits, the plane could go from carrying 8,000 pounds of weapons to a 12 passenger transport, and darn near anything in between!

Its basic weapon load was four 20-mm cannons and 8,000 pounds of bombs. During Vietnam, it was not unheard of to see Skyraiders sporting up to eight 20-mm cannons!

Douglas A-4 Skyhawk (USA)
While the A-4 was originally designed to carry a single weapon on a single type of mission — nuclear bombs, it quickly became a favorite among pilots in Vietnam and in Air Forces around the world. Universally called the Scooter, this lightweight, compact attack bomber was a real sports car for a bomber. It was agile, responsive and pilot friendly. At least 10 different versions were built by McDonnell Douglas, with the A-4C, A-4E and A-4F versions accounting for about half of the 2,300+ jets produced. The versions flown by the U.S forces

A time exposure of the work of a AC-47. It is said that a two second burst from Spooky could put a round in every square foot of a football field. (USAF Photo)

An A-1 Skyraider loaded for action. The 500 lbs bombs sport pipe extensions to ensure an above-ground detonation.

Closeup of the bomb load of a Skyraider. The wing leading edge conceals two 20-mm cannons on each side, for even more firepower. (Vadnais Photo)

A-7E Corsair II attack aircraft. Earlier versions of the A-7 flew thousands of combat missions in Vietnam from carriers off the coast. (US Navy Photo)

A close up view of the six barreled Gatling gun carried by the Corsair II. (Vadnais Photo)

came with two 20-mm Mk 12 cannons in the wing root, four underwing hardpoints and a fuselage mounting point that could hold a variety of weapons or stores. Some of the A-4s manufactured, re-manufactured, for foreign military sales included 30-mm cannons in place of the 20-mm cannons.

Grumman A-6 Intruder (USA)

Built in large numbers strictly as an attack bomber, the Intruder never carried a gun or cannon. A capable aircraft in its own right, it carries with it the "honor" of being shot down more than any other aircraft during the Vietnam War.

LTV A-7 Corsair (USA)

Designed to become the backbone of carrier light attack aircraft, the Corsair was intended to be subsonic but with long range and endurance as performance priorities. It was to re-

The prototype, the YA-37, armed with one GAU-2/1 7.62-mm Gatling gun. Notice that five of the six barrels are covered by a yellow cover, leaving only the barrel firing uncovered. Rationale for this is undetermined. (Vadnais Photo)

A-37 from the 8th Tactical Fighter Wing armed with bombs, headed North. (USAF Photo)

place the A-4 Skyhawk in this attack role. It could carry a wide variety of weapons, and specialized in close air support and battlefield interdiction. The A-7A and A-7B models carried a two Mk 12 20-mm cannons with 680 rounds, but that was upgraded in the -C model with a single M61A1 Vulcan cannon with 1,000 rounds. Eventually, 535 Corsairs were built.

Cessna A-37 Dragonfly (USA)
In the early 1960s, the U.S. military developed a number of existing aircraft for counterinsurgency warfare. Re-engined with new General Electric engines, the A-37 had more than twice the power of the T-37. With reinforced structures, the Dragonfly was a cheap and capable platform for its role of fighting the low intensity conflict mission. Armed with one 7.62-mm GAU-2B/A minigun in the nose. It also had eight hardpoints capable of carrying a total of about 5,000 pounds.

Hawker Siddley/McDonell Douglas AV-8 Harrier (USA/British)
The Marines have long lusted for a fighter capable of giving close air support to Marines landing on a beach. However, for longest time technology prevented this from happening. Then the Harrier came along. They spent months evaluating the plane in 1968, while it was still in development, and found it to be the answer to their prayers. Capable of taking off and landing vertically, Harriers didn't need big carriers to operate from, and could closely follow the Marines landing in a hostile area. Upgraded constantly since introduced in 1971, the Harrier is still a main part of Marine Corps battle doctrine. It is

F-100D being readied for a mission to North Vietnam. (USAF Photo)

A Thal Security Police member stands guard over an F-100D deployed to Thailand for operational testing. (USAF Photo)

Three F-100Ds take on fuel from a KB-50 tanker. (USAF Photo)

Weapons spread for an F-104 Starfighter – a pretty impressive sight for a plane with minimal wing area for attaching weapons. (USAF Photo)

armed with two 30-mm cannons of British manufacture, each cannon has a supply of 130 rounds. The Harrier also carriers two AIM-9 Sidewinders and up to 5,000 pounds of bombs, missiles or drop tanks.

North American F-100 Super Sabre (USA)

An evolution of the F-86 Sabre, the F-100 was originally armed with four M39E 20-mm cannons with an APX-6 radar gunsight, the Super Sabre was the Air Force's first fighter capable of supersonic speed in level flight. Later versions of the Super Sabre dropped the M39E cannon and instead carried four T160 20-mm cannons. It also had four hard points capable of carrying up to 7,500 pounds of weapons, including nuclear weapons.

The Super Sabre's began leaving the Air Force inventory in 1958 and were either stored at Nellis AFB, Nev., or sold to foreign air forces, including Nationalist China. Later versions of the jet, the F-100C, saw action in Vietnam. In 1965, the first "Wild Weasel" squadron – of F-100Fs – deployed to Thailand for combat operations.

McDonnell F-101 Voodoo (USA)

The basic development began in 1946 to escort bombers of the newly created Strategic Air Command. It wasn't until the early '50s that production began though. Although it began life as a bomber escort, the Voodoo excelled as both an interceptor and as a supersonic reconnaissance platform.

The RF-101 flew its first combat missions in Vietnam during 1961, and did most of the tactical reconnaissance until 1965 when the RF-4 became operational. All versions of the Voodoo began leaving the U.S. Air Force in 1968 and all were either retired, sold or transferred to the Guard by 1971.

The F-101A/C versions were armed with four M-39 20-mm guns while the F-101B/F versions did not carry guns only two AIR-2A rockets and two AIR-4C missiles. The RF-101 versions were unarmed.

An F-104C being readied for a night mission in Vietnam. (USAF Photo)

"Feeding the beast" belts of ammunition, ground crewmen ready an F-105 Thunderchief. (USAF Photo)

Close up view of the M61 Vulcan 20-mm cannon carried on the F-105D. (Vadnais Photo)

F-105 ready for action. Note the opening for the M61 near the nose of the aircraft. The cannon ejected the spent shells from the bottom of the aircraft and they often chipped away the paint on the bottom of the jet. (USAF Photo)

Convair F-102 Delta Dagger (USA)

The Air Forces first fighter to be principally armed with guided missiles, the Dagger was also the first supersonic all-weather interceptor. It was also the first delta winged fighter to see operational use.

The F-102 were deployed to Vietnam for air defense originally, but later they were used to escort the B-52s on their missions. During their nearly 10 year-long war presence, the Dagger had an enviable record – just 15 were lost to enemy fire

The basic armament on the F-102 were 12 2.75-inch Folding Fin Aerial Rockets and a variety of air-to-air missiles, but no recorded use of guns or gun pods.

Lockheed F-104 Starfighter (USA)

Designed as a lightweight air superiority day fighter to replace the F-100, the Starfighter was an unsolicited proposal from Lockheed. The Starfighter was one of the Air Force's smallest and lightest planes. As such, it is hard to image its development into a nuclear capable fighter bomber with air to ground capability.

The original design had two 30-mm guns but after the original mockup was built, that was changed to one 20-mm M-61 Gatling gun currently underdevelopment. This resulted in an 80 pound weight drop.

The F-104C first went to Vietnam in 1965 on a temporary basis. One squadron stood alert at Kung Kuan, Taiwan and at Da Nang, South Vietnam. From Da Nang, the Starfighter hit targets in both South and North Vietnam. The Starfighter returned to the theater in 1966, on a permanent basis, assigned to Udorn Royal Thai Air Force Base. They were replaced a year later by the F-4 Phantom.

Republic F-105 Thunderchief (USA)

In developing the Thunderchief, Republic saw it as the successor of the F-84. The Air Staff agreed, and ordered the

Ground crew checking the 20-mm Vulcan cannon on the Thunderchief. The red star below the cockpit denotes a North Vietnamese kill. (USAF Photo)

The F-106 Delta Dart as a front line fighter, undergoing tests at Edwards AFB, Calif. (USAF Photo)

aircraft to production without considering any alternative designs.

Significant growing pains continually delayed the program, and once it entered operational service, the F-105 had abnormally low availability rates. As a matter of fact, it required 150 man-hours of maintenance for every flight hour!

Despite all these teething problems, the Air Force decided to use the Thunderchief as a replacement for the F-100Cs used by the Thunderbirds, the Air Force Aerial Demonstration Team. The team converted in 1963/1964, with the last of nine aircraft being delivered 10 days before the first scheduled performance. A month later, a serious accident, and subsequent modifications required to the airplane forced the Thunderbirds into "temporarily" going back to the F-100s, where they stayed until 1969 when the team converted to the F-4E.

F-105Ds, flying from Korat Air Base, Thailand, began operations in Southeast Asia in early 1965. In the years that followed, they carried out more strikes against the North than

Now used as target aircraft, a QF-106 departs Tyndall AFB, Fla., on a manned mission. The plane can also be flown by remote control, when missiles are fired. Manned missions are either currency training or dissimilar aircraft air combat training. (Vadnais Photo)

Initially not equipped with a gun system, later versions of the F-4 Phantom would have them installed for combat missions. (USAF Photo)

The F-106 was not designed with a gun built into it, but this one had one added for test and evaluation. The M61 Vulcan 20-mm cannon is located along the bottom of the fuselage, where a center line fuel tank would normally be seen. (USAF Photo)

The F-111 had a troubled deployment to Vietnam, but it later came into its own as a front line fighter in NATO. Shown here is an FB-111, a fighter-bomber version that was used during Operation EL DORADO CANYON in 1989. (USAF Photo)

any other Air Force aircraft. The airplane continued to receive major modifications in the field, including armor plating, backup flight controls, and new gun/bomb sights. Through all this, the airplane began to get a reputation as a great war plane. It was praised for its payload, range and exceptional speed at low altitudes.

The Thunderchief carried a single M-61 20-mm gun with 1,029 rounds and could carry a wide variety of bombs and air-to-ground missiles.

Convair F-106 Delta Dart (USA)

Developed as the "ultimate interceptor," the Delta Dart was an improvement on Convair's F-102. It continued the delta wing planform, but had significant changes internally. The Dart was designed from day one to not have a gun, only missiles.

Built as an interceptor for Air Defense Command, the Delta Dart did not deploy to Vietnam during the war. However, it did deploy to Korea in 1968 during the tensions arising from the sinking of the USS Pueblo by the North Koreans.

Still in service after nearly 40 years, now designated the QF-106, the airplane is used as aerial targets by the 475th Weapons Evaluations Group, Tyndall AFB. The last QF-106 mission flew in January 1998, and then all Delta Darts were retired. The aerial target role now being flown by the QF-4.

General Dynamics F-111 Aardvark (USA)

The F-111, in its original fighter configuration, had an operational requirement to reach Mach 2.5, 60,000 feet, all-weather STOL fighter. It was supposed to take off and land on grass fields in less than 3,000 feet, with an 800-mile combat radius, carrying either conventional or nuclear weapons – pretty challenging!

All fighter versions carried a single M-61A1 Vulcan 20-mm Gatling gun.

In March 1966 Air Force leaders decided to rush a small number of F-111As to Vietnam to both boost night and all-weather attacks and test the aircraft's combat capability. As is the case most times in rushing things, it was not a good idea. The project, called Combat Lancer, was not a success.

An F-4C assigned to Luke AFB, Ariz., over the Southwestern United States. (USAF Photo)

The newer version of the gun pod, the GPU-5/A 30-mm cannon. This pod fits on both the Phantom and the F-16 Falcon. (USAF Photo)

Rearming and reloading an F-5 fighter in Vietnam. (USAF Photo)

After one month and 55 combat missions, four of the six deployed Aarvarks had been lost. Replacement aircraft were sent, but lose of another aircraft stopped the operation. It would take another six years before the plane would return to the theater.

The return, in 1972 was a success, but just barely. Four F-111s could deliver as much bombs as 20 F-4, in weather that would ground all other aircraft. While this was a significant cost saving, the F-111 had problems that took awhile to overcome. Continuing problems with engine stalls in heavy rain (of major concern for an all-weather fighter), problems with the Terrain Following Radar and attack radar kept the F-111 from being an unqualified success. Seven of the 52 F-111As deployed this time around were lost while flying more than 3,000 combat missions.

McDonnell F-4 Phantom (USA)

In its original configuration, the Phantom did not carry a gun. The school of thought in the Air Force then was geared towards fighting a missile engagement, and the gun was unnecessary weight. With the advent of the F-4C, a gun pod was available, and when the F-4E rolled off the production line, a M61A1 20-mm six barrel cannon was mounted permanently in the nose. Every aircraft since then has included a gun in the basic design. Who says the Air Force doesn't learn from its mistakes?

The Phantom was, and continues to be, an awesome piece of machinery. It performs a wide variety of missions ranging from interceptor, nuclear strike, anti shipping and suppression of enemy air defense. Along with the U.S., large numbers of F-4s flew (and continue to fly) in the air forces of

A B-52 takes off nose low from RAF Mildenhall during its annual air show, Air Fete. (Vadnais Photo)

The tail stinger of a B-58 Hustler—a 20-mm cannon mounted in the tail. (Vadnais Photo)

Business end of a MiG-19, with its three NR-30 30-mm cannons. (Vadnais Photo)

South Korea, Australia, Israel, Turkey, Japan, Egypt, Iran, Greece and Germany.

Northrop F-5E Tiger II (USA)
Chosen as the winner of the International Fighter Aircraft competition in 1970, the Northrop F-5E was an updated version of the F-5A. It's design and construction emphasized aerial agility over pure speed. Armed with two 20-mm cannons, with 280 rounds each, in the nose, the Tiger II served as a tactical fighter in many countries and as the aggressor in dissimilar air combat maneuvers.

Mikoyan-Gurevich MiG-17 (Soviet Union)
A carry over from the Korea era, the MiG-17 was totally outclassed by the American hardware. Basically, it was an advanced version of the MiG-15 with an afterburner and a redesigned wing. It was armed with one 37-mm cannon and two 23-mm cannons in the lower nose area. Later versions could carry two air-to-air missiles, but early versions were gun and rocket only.

American pilots downed 61 MiG-17s during the war, with F-105 pilots notching 27 kills and F-4 Phantoms adding 33 more. One kill was shared by an F-4 and an F-105.

Mikoyan-Gurevich MiG-19 (Soviet Union)
Based loosely on the F-86, the MiG-19 spent many years fighting the air war. Unfortunately, it didn't fare very well. It was armed with three 30-mm NR-30 cannons, two air-to-air missiles and rocket or 37-mm cannon pods.

A total of 8 MiG-19 were downed by American pilots, all by different variants of the F-4 Phantom.

Mikoyan-Gurevich MiG-21 (Soviet Union)
This high speed air superiority fighter could top Mach 2 at altitude. It's speed was its primary defense against American pilots, it would enter a fight, fire off a couple of missiles and beat feet to the north.

It didn't carry much firepower, only one 37-mm cannon in the nose and a 30-mm cannon in the along the fuselage. It also carried two air-to-air missiles

Test flight of the XB-70 Valkyrie in 1965. It didn't carry any guns (USAF Photo)

The NR-30 30-mm cannon on a MiG-21F. (Vadnais Photo)

Though pretty simple in design, what it did do, it did very well. It took on the newest, best American equipment and held its own. There wasn't a single American pilot who took the MiG-21 lightly. Because it was so widely used by the North Vietnamese, it has the distinction of being the most shot down aircraft in Vietnam—68 times a MiG-21 fell to American pilots. Two of the kills were by B-52 tail gunners.

Boeing B-52 Stratofortress (USA)
Still a world champion bomber, the BUFF continually has evolved to meet the needs of the Air Force. Originally designed to carry nuclear bombs to the USSR, it now is a conventional bomber, a cruise missile carrier, and a mine-laying aircraft. From its first flight until 1994, it carried a tail stinger gun, either four .50-cal machine guns or two 20-mm M-61A1 Gatling guns.

During the Vietnam war, it flew more than 126,600 combat missions, bombing targets ranging from bike paths to real, operational targets during Rolling Thunder and Operation Linebacker I and II.

Still, as a bomber, it had two aerial kills during the war. The first bomber kill came on Dec. 18, 1972 during Operation Linebacker II. Staff Sgt Samuel O. Turner was the tail gunner on a B-52D, hitting targets in the Hanoi area. His kill of a MiG-21 was witnessed by another gunner, Master Sgt Lewis E. LeBlanc. Turner received a Silver Star medal from Gen John C. Meyer, commander-in-chief, Strategic Air Command, for his shootdown. The only other B-52 kill came from Airman First Class Albert E. Moore, on Dec. 24, 1972.

Altogether, the Air Force lost 31 B-52s during the war – 18 to enemy fire (all over North Vietnam) and 13 from other causes. (Additional data contained in the Modern Gun Era chapter.)

Boeing B-47 Stratojet (USA)
The cutting edge Stratojet was the first swept wing bomber and used a tandem bicycle landing gear (similar to the U-2) that resulted in a very thin wing. Its development began before the Korean War, but because of numerous problems in the development and test, it never made it to the war. Nearly 2,000 were eventually produced in a variety of variations.

Martin/General Dynamics B-57 Canberra (USA)
The only bomber in Air Force operational service that wasn't of American design. This light tactical bomber entered U.S. Air Force service in 1955, but many were quickly modified into high altitude reconnaissance aircraft, becoming the RB-57D. Later, the B-57E became the first purpose built target tug. The Canberra did see action later on, in Vietnam, for some daylight bombing, night interdiction and reconnaissance missions. It was generally considered to be an average aircraft platform for bombing, but did much better in its reconnaissance and electronic surveillance roles.

Convair B-58 Hustler (USA)
The Hustler was the first supersonic bomber in Air Force operation, and became the first bomber to reach Mach 2. The unique design of the Hustler featured many innovations. Among them were the honeycombed internal structures to reduce weight, individual escape capsules for each of the four crewmembers and innovative star tracking navigation techniques. The weapons load was carried external to the plane, with the nuclear bomb being nestled into the centerline fuel tank. The B-58 set more world records than any other combat aircraft in the world, including the Los Angeles-to-Washington D.C. record (2:00:56, average speed 1,214.71 seconds) since broken by the SR-71. The Hustler had a dubious record of crashing, totaling 26 crashes out of the 116 aircraft built – not the best safety record!

Douglas B-66 Destroyer (USA)
The B-66 was used by the Air Force as tactical light bomber and a reconnaissance aircraft. The Destroyer carried two remotely controlled 20-mm cannons in the tail turret, and 15,000 pounds of bombs. The B-66 overcame early problems that nearly canceled the program to become a jack of all trades aircraft. The B-66B was the only bomber-only version built. Other versions included the RB-66 (for reconnaissance), the

Loading an SUU-23/A gun pod at Phu Cat Air Base, Vietnam. (USAF Photo)

EB-66 (electronic warfare) and WB-66 (battlefield weather data collection).

In Vietnam, the EB-66 was the most utilized version. They were used to locate North Vietnamese radar sites, determine their function and identify frequencies. These aircraft were eventually retired in the early 1970s because of their extremely heavy use during the war. Their role was filled by the EF-111.

North American B-70 Valkyrie (USA)
The experimental XB-70 was designed and built to be a supersonic strategic bomber. The Valkyrie first flew on September 21, 1964. The program faced many delays due the futurist design and cutting edge technology requirements. The second Valkyrie flew in 1966 and reached Mach 3 on its maiden flight! Later that year, this second aircraft had a mid-air collision with an F-104 during a photo shoot and both aircraft crashed with all on board killed. The program was canceled shortly thereafter and the sole remaining XB-70 was flown to the Air Force Museum in Dayton, Ohio.

Grumman OV-1 Mohawk (USA)
A tactical reconnaissance aircraft, a small number of OV-1As were modified during the Vietnam war with the addition of six underwing hardpoints for hanging bombs, rockets or gun pods for close air support. The vast majority of the Mohawks were equipped with cameras and side looking airborne radars and other sensors for detecting enemy radar systems.

Rockwell International OV-10 Bronco (USA)
Designed and built in the early 1960s, the Bronco filled a U.S. Marine Corps need for a counter insurgency aircraft. The Marines flew them in the counter insurgency role, and helicopter escort while the Air Force bought more than 150 for the forward air control role in Vietnam. Each boom housed a 7.62-mm M60C machine gun with 500 rounds. The five hard points on the aircraft could carry a variety of weapons, ranging from bombs and rockets to machine gun and cannon pods.

Captain's Steve Ritchie and Charles DeBellevue, two aces of the Vietnam war. (USAF Photo)

A newer version, called the OV-10D was delivered to the Marines in the late 1970s. It was specially designed for night observation surveillance. Equipped with FLIR sensors, laser designators and armed with an M197 20-mm, three barreled cannon with 1,500 rounds, the new Bronco was a formidable lightweight gunship.

The Vietnam Aces:

Charles D. DeBellevue, Capt, USAF, 6 victories
Randy Cunningham, Lt, USN, 5 victories
William Driscoll, Lt, USN, 5 victories
Jeffrey S. Feinstein, Capt, USAF, 5 victories
Richard S. Ritchie, Capt, USAF, 5 victories

Chapter 5:
The Modern Gun Era

In the modern era, the gun's emphasis seemed to be fading, at least that was one way to look at it. A large part of the defense for the modern fighter was being taken over by air-to-air missiles, and the air-to-ground mission with air-to-ground missiles.

But even though the modern aircraft would only be carrying one gun most of the time, the capability of these modern Gatling Guns greatly superseded the capability of the multiple guns of the past. There was also a minimal use of externally-mounted gun pods during the time period.

The era of the gun, though, with modern bombers appears to be over. The latest two USAF bombers, the B-1 and B-2, would have NO guns.

FIGHTERS

Northrop F-5 Tiger (USA)

One of the most-heavily produced fighters on the international markets, the F-5 retained a heavy gun protection arrangement with a pair of 20-mm cannons located in the nose. The Tiger has been around since 1972 and continues to serve with a number of South American, Africa and Far East Air Forces.

Grumman F-14 Tomcat (USA)

The single gun theory was used on the awesome F-14 Tomcat with its location being on the port side of the lower fuselage directly below the cockpit. The General Electric M61 multi-barrel Gatling Gun was augmented by short range air-to-air missiles providing for excellent defensive capabilities. The Tomcat first flew in 1970 and was the top fighter for the US Navy until the appearance of the F-18 Hornet in the 1980s.

Here's a view of the super-potent M-61 gun removed from its normal F-15 location. As can be seen, this is no small weapon system. (GE Photo)

Here's a unique F-16A of the Syracuse Air National Guard fulfilling the CAS (Close Air Support) mission. Note the 30-mm gun pod mounted underneath the fuselage. The gun game is NOT dead with modern fighters. (USAF Photo)

The fact that the F-15 Eagle carried a single 20-mm M-61 multi-barrel gun with 960 rounds on board was pretty much status quo for such modern fighter aircraft. What is strange, though, is where the gun is located, in the upper engine housing of the right engine. (McDonnell Douglas Photo)

McDonnell Douglas F-15 Eagle (USA)

Missiles and gun were the weapons suite for the Air Force's top fighter in the 1980s, 1990s, and into the next century. First flying in 1972, the Eagle starred in Desert Storm, and as of 1996, had a perfect record in air-to-air combat, i.e. not a single Eagle lost to enemy aircraft.

AIM-7, AIM-9, and AIM-12D air-to-air missiles augmented the standard M61 Gatling Gun which carried almost a thousand rounds. Unlike the fuselage mounting location of the F-14, the Eagle carried its M61 in the top of the right-side engine pod. The ammunition feed came from the ammunition barrel located in mid-fuselage directly behind the cockpit.

The F-15 was born from the FX program in the late 1960s, and the need for the fighter came from the Soviet MiG-23 and 25 which were challenging US aerial superiority.

General Dynamics F-16 Fighting Falcon (USA)

Gazing at the F-16 from the left side, it's easy to see the location of the again-expected M61 Gatling Gun. The gun is located directly below the cockpit. The feedbelt connects the rear of the gun to the ammunition drum directly aft of the cockpit. Again, the gun capability of the Falcon is augmented by air-to-air missiles.

In the late 1980s, an interesting experiment was accomplished using F-16As of the New York Air National Guard when the model was selected for modification to a CAS(Close Air Support) configuration. It really brought back the importance of the aircraft gun as the planes carried General Electric GPU-5 30-mm gun pod slung underneath the fuselage.

The experiment was carried further when it was decided later to modify a number of F-16C/D models into the CAS configuration. In addition to the 30-mm gun pod, other additions would include an improved data line, digital terrain mapping system, Pave Penny laser spot tracker and a forward-looking infrared(FLIR) system for night operations.

McDonnell Douglas F/A-18 Hornet (USA)

For the Navy's front-line fighter, the single M61 Vulcan was retained, however the location was changed from its location in the USAF F-15 and F-16. The weapon was located in the forward fuselage with the barrels just barely protruding in the nose cone.

Even though it carries an 'F' fighter designator, amazingly, the F-117 does not carry a single gun or missile, only smart bombs. Maybe it should have been called a bomber? (USAF Photo)

The protrubance is so slight that it hardly effects the aerodynamics of the plane. All versions of the Hornet carry the M61, even the latest F/A-18E/F version.

Lockheed-Martin F-22 (USA)
The latest USAF fighter, the F-22, will also have an M-61 Gatling Gun although there were reportedly discussions that considered it being deleted. The gun will be carried in a forward fuselage location.

Lockheed F-117 Nighthawk (USA)
It's interesting that the super-secret stealth fighter even got a fighter designation since it has guns or missiles, but performs more like a bomber with its bomb load.

There is no problem identifying the F/A-18's location of its onboard gun system, right there on the forward nose. The weapon is the M61 Gatling, the same weapon used on the Air Force F-15.(McDonnell Douglas Photo)

It was reported that there were considerations for eliminating the gun in the new F-22 design. But that didn't happen and the stealthy new fighter carries the expected Gatling Gun.

But then again, when you can swoop onto a target, drop your ordnance, and then escape before the enemy even knows you were there.

Dassault Rafale (France)
One of the mainstay fighters of the French Air Force is the long-standing Dassault Rafale. The gun is in place with one 30-mm DEFA 791B gun located in the engine duct. The gun is augmented in a big way with air-to-air missiles located on underwing MICA air-to-air missiles.

Panavia MRCA Tornado (British, German, Italy Consortium)
Any mention of modern fighter aircraft, mention must be made of the Tornado, a fighter which has been around since its first flight in 1974. The swing-wing fighter is capable of carrying some 15,000 pound of ordnance carried on three fixed body pylons and four swiveling wing pylons.

The gun armament consists of a 27-mm Mauser cannon located in the lower fuselage, augmented by a suite of Sidewinder air-to-air missiles.

Mirage 2000 (France)
Long the primary combat fighter of the French Air Force, the Mirage 2000 was initially developed as an interceptor. A number of different versions have evolved through the years, but all versions have retained the initial gun installation of a pair of 30-mm DEFA 554 machine guns located in the lower fuselage.

The MiG-29 is one of the top Russian fighters, bearing a marked similarity to U.S. designs. There is still a gun on-board with a 30-mm cannon. (Air Force Museum Photo)

From the beginning, the A-10 was designed around its 30-mm Gatling Gun. This is an early factory concept drawing. (Fairchild Photo)

A-10 INBOARD PROFILE

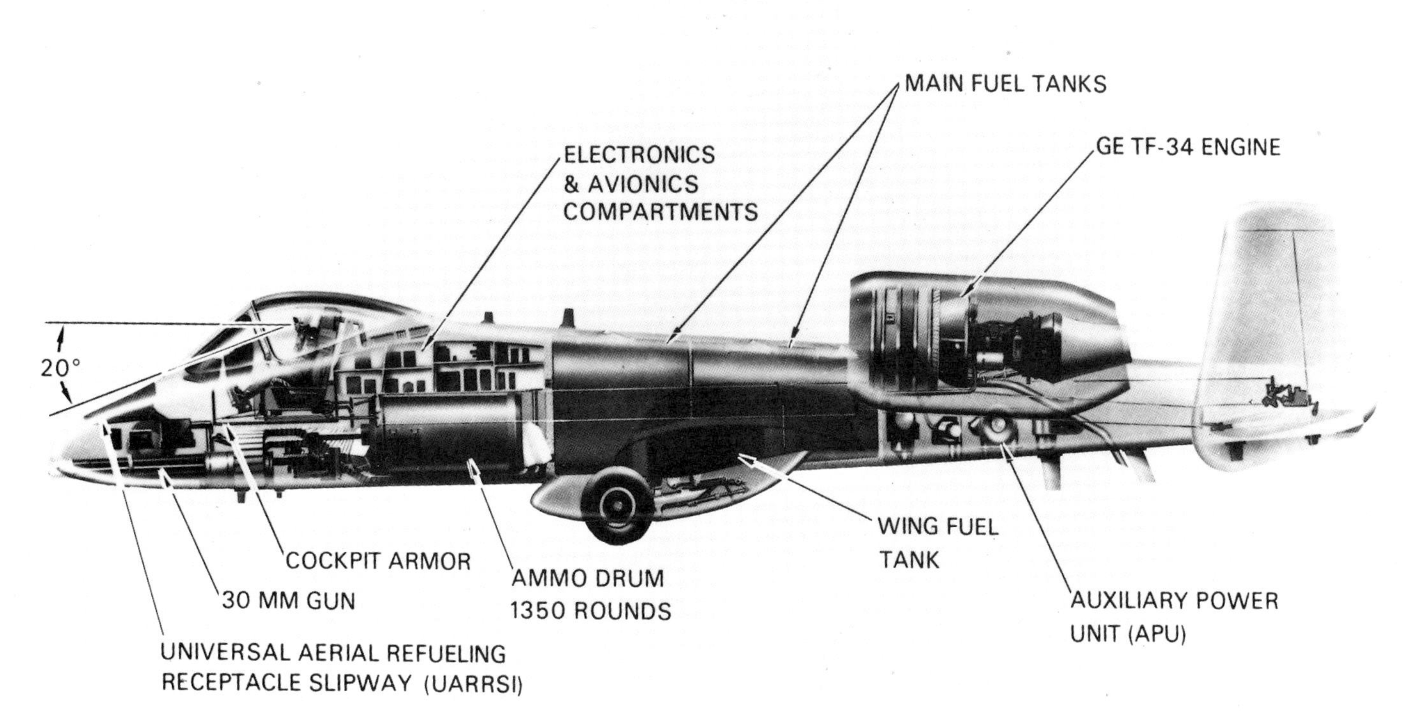

This cutaway drawing shows just how much of this close support aircraft is donated to its 30-mm gun system. Notice how much volume the ammo barrel takes up. (Fairchild Photo)

A view of the A-10 Ammunition Loading System, called the "Hydra". (Fairchild Photo)

Consolidated Eurofighter 2000 (France, Germany, Spain, Italy, UK)

Even with this new consortium fighter for the next century, the gun is still playing a role in aircraft defense. Granted, there is only one gun as compared to a number of air-to-air missiles, but planners still felt its need.

In the case of this futuristic fighter, which will be fielded in the next century, it will carry one fuselage-mounted 27-mm Mauser cannon

Saab AJ37 Viggen (Sweden)

Carrying one of the most unique airframe designs, the Swedish Viggen has been a longtime mainstay with the Swedish Air Force.

Its design, with large engine intakes located far forward, isn't compatible with nose mounted guns, hence the possible reason for going to use of gun pods. The weapon of choice is a British Aden 30-mm cannon with 150 or 200 rounds.

Primarily, though, the defense of the aircraft was more heavily aligned to air-to-air missiles.

Sukhoi Su-24 Fencer (Russia)

The swing-wing Fencer is one of the mainstays of the Soviet Air Force and continues to use a gun, in addition to air-to-air missiles for defense and offense.

Carrying on the international trend of Gatling-style machine guns, the Fencer uses a 30-mm six-barrel Gatling which is mounted in a fairing on the starboard side of the forward fuselage. Also, there is a similar fair on the port side which probably houses the ammunition feed system.

The next Sukhoi model, the Su-27 carries a 23-mm cannon in addition to its expected suite of missiles.

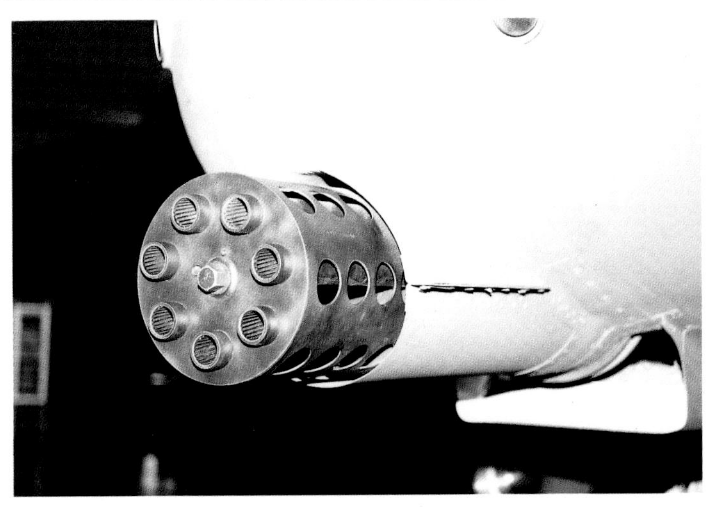

The business end of the GAU-8 gun system clearly showing the barrel ends of the seven guns that comprise the system. (Vadnais Photo)

Mikoyan MiG-27 Flogger (Russia)

The gun remained in place with the Flogger with initial versions carrying the twin-barrel Gsh-23L twin-barrel gun. Later

This A-10 front end photo shows the extremely low-mounting position of the GAU-8 Gatling Gun. Note the exhaust duct directly under the barrel. (Bob Shenberger Photo)

When the GAU-8 gun is fired, the effect on the A-10 aircraft structure is definitely noticed. It has been reported that the sizable recoil of the firing can actually slightly slow the forward velocity of the plane. (USAF Photo)

versions incorporated a new-for-the-time 23-mm six-barrel rotary cannon.

Both versions were carried in a location aft of the front landing gear on the fuselage underside. That latter gun was more effective against ground targets with some 700 rounds carried onboard.

In addition, the gun installation was augmented with the expected air-to-air missile systems.

Mikoyan MiG-29 (Russia)
One of the most advanced fighters from the former Soviet Union is the twin-tailed Mikoyan MiG-29. The gun is still in place here with a 30-mm GHs-301 cannon which is carried at the juncture of the left wing leading edge and lower fuselage.

Mikoyan MiG-31 (Russia)
One of the latest versions from the Mikoyan Design Bureau, a design that looks amazingly like the USAF F-15 incorporates a gun with the 23-mm multi-barrel GSh-6-23 cannon.

PREVIOUS and LEFT: This sequence of eight photos shows the effects of the 30-mm GAU-8 A-10 gun system during a test on US M-48 tanks. The fireballs seen coming from the tank are secondary explosions caused by the rounds penetrating and igniting internal and fuel stores in the tank. (Fairchild Photo)

ATTACK AIRCRAFT

Fairchild A-10 Warthog (USA)

In an era when the importance of the gun seemed to be decreasing, the A-10 was a plane that was designed from scratch to be a platform for a gun. The plane was the first Air Force aircraft to be developed to deliver aerial firepower to defeat the potential enemy ground threat.

The centerpiece of this ground attack machine was the awesome GAU-8/A 30-mm Gatling gun system mounted internally along the aircraft's centerline. The system has proved itself far more than was ever expected and continues to be a viable weapon system in the 1990s.

The A-10 proved itself during the war in Iraq when it destroyed a wide array of ground weapons including tanks.

An impressive array of A-10 offensive firepower highlighted by the massive array of 30-mm rounds for the GAU-8 gun. (Fairchild Photo)

The size of the A-10 is vividly illustrated by this comparison with a VW Beetle. You have to be kidding! (USAF Photo)

A view of the tail guns of the B-52C shows the four barrels that was the characteristic of the Stratofortress through the G model. The final H version reverted to a single barrelled gun that would finally be removed in the mid-1990s. ***See B-52 section p. 73.*** **(USAF Photo)**

This is a ground view of the tail-stinger of the many-produced B-52D. This particular plane is on display at the Air Force Museum. ***See B-52 section p. 73.*** **(Vadnais Photo)**

Of course, the aircraft carries a combination of other attack weapons including conventional and laser guided weapons, rockets, cluster bomb munitions, and Maverick missiles.

But again, it's that magnificent gun that sets this plane apart. In going after a ground target, the A-10 has an extremely high probability of a kill on the first pass. In addition to the armor-piercing projectile, which is capable of penetrating medium and heavy armor, the guns also fires high explosive incendiary ammunition which is extremely effective against a wide variety of softer targets such as trucks, armored personnel carrier, and other vehicles. The maneuverability of the A-10, combined with the accuracy of the gun, provides a system for engaging targets even in adverse weather conditions.

The gun itself fires at rates of 2,100 or 4,200 rounds per minute. The weapon is a conventional Gatling-type gun with seven barrels and is fed ammunition from a linkless ammunition feed and storage system holding 1,350 rounds. The revolutionary Ammunition Loader System was a major contributor in reducing aircraft turnaround time from three hours to about 15 minutes.

With its outstanding close support effort in Iraq, the A-10 lengthened its lifetime indefinitely. Originally planned for retirement in 1994, the A-10 stands in the mid-1990s as a plane that is always given first consideration in any third world conflict.

And although the A-10 has not always been given the best reviews because of its relatively slow speed capability(would you believe "Warthog"??), the A-10 recently proved its worth by winning the Air Force Gunsmoke bombing competition against the glamour F-15 and F-16 fighters.

It could well be the last manned bomber. It's an amazing offensive platform, the B-2, but one weapon system which was never considered was a gun or cannon. (USAF Photo)

Conceived in the early 1970s, the B-1 Bomber was never designed with a gun and it remains that way today. (USAF Photo)

In addition, a recent adaptation to the A-10 system, called the Low Altitude Safety and Targeting Enhancements LASTE) to improve the effectiveness of the gun during air-to-air missions because it compensates for the A-10's downward recoil after cannon fire.

It was a plane built for a gun and it's a concept that will not be repeated.

VTOL-STOL AIRCRAFT

McDonnell Douglas/Bae AV-8B Harrier II (USA)

This short take-off aircraft had great capability for carrying ordnance from a number of underwing pylons. But its own protection wasn't forgotten with the adoption of a 25-mm cannon based on the General Electric GAU-12/U with three hundred rounds and mounted under the fuselage.

Bell/Boeing Vertal V-22 Osprey (USA)

Even though it's not exactly a helicopter, the V-22 Osprey performs just like one and is therefore included in this section. The newest Army/Marine system will provide the role of an assault transport possessing a VTOL capability with its pivotal engines.

The only armament on the V-22 comes from the old reliable gun in the form of a forward-firing nose-mounted 12.7-mm(0.5inch) multi-barrel machine gun. The gun's barrel sticks out far forward of the craft's nose.

It's a brand-new concept, but still relying on the age-old gun armament technique.

BOMBERS

Boeing B-52 Stratofortress (USA)

It's hard to consider the B-52 bomber in a modern connotation, but since later versions of the eight engine bomber will be flying well into the 21st century, it deserves the honor even though its roots stretch back to the 1950s.

Following the era of heavily-armed bombers, all versions of the B-52 would carry only one gun installation, its location being the rear of the aircraft.

A number of different gun systems were carried in different B-52 models including the following: American Bosch A-3A system (four .50-cal guns in hydraulically-driven turret, gunner in tail)-B-52A/B/C/D, MD-5 system(four 20mm cannons in electrically driven turret, gunner in tail)-B-52B, MD-9 system(similar to A-3A)-B-52C/D/E/F, Avco-Crosley AN/ASG-15 system(four .50-cal guns, gunner forward)-B-52G, and Emerson Electric AN/ASG-21 system(one 20-mm M-61 Gatling Gun, dual search and track radars)-B-52H.

Maybe it was a sign of the times, but in the mid-1990s, it was decided by the Air Force to remove the rear Gatling Gun installation from the still-operational B-52H models.

It was hard to figure the reasoning since the bomber still will be performing similar missions that it had done in the past, but that was the decision that was made.

Maintenance crews covered the hole left by the gun installation and added weight to maintain aircraft balance while in flight.

ABOVE, BELOW, and ACROSS: A number of recent programs have investigated the phenomena of high angle of attack(AOA) flight which could aid a fighter pilot while using a gun or cannon. Such an effort was the forward-sweep X-29 program. (USAF Photo)

Rockwell International B-1B (USA)

A brand-new bomber for the 1980s and beyond, and a brand new philosophy about onboard defense systems. There were none. No air-to-air missiles, no rockets, and for purposes of this book, NO GUNS. They were never even considered.

It was a different world from World War II when bombers bristled with machine guns and cannons. No longer as planners figured the missions the B-1B would perform would not require that type of protection.

The bomber was planned to carry either air-launched missiles or bombs, and that was the only ordnance which was onboard. The airborne gun had been forgotten on the high-tech bomber.

Northrop B-2 Spirit (USA)

Possibly the final manned bomber, the futuristic stealth-possessing B-2 followed the trend set by the B-1 carrying no guns of any type. It's probably a trend that will continue, but with any third world war, there may be regret that no guns exist on these later bomber aircraft.

Tupolev Tu-26 Backfire (Russia)

Entering service in the mid-1970s, the supersonic Backfire presented real challenges to Allied defenses. The mission was that of a long-range bomber and a missile-launching platform.

With those missions, the decision was made to omit any gun installations, the same decision that was made with latter US bombers.

BELOW: A Bell-Boeing V-22 is shown with engines in vertical position during a test takeoff. (Bell Photo)

Chapter 6:
Gunships, Fixed Wing, & Helicopter Gunships

Just when you think that they have thought of everything, somebody comes up with an idea so off-the-wall that after the first burst of laughter, the thought "now wait a minute...that might actually work!" runs through your mind, as the eyes light up.

The concept of the gunship is not new. In fact, many of the principles involved, side shooting weapons, the "pylon" turns that keep the target continuously in the gunsight, date back to World War I. During the latter years of WW II, Gilmour C. MacDonald, an Air Force Lt Col., conceived of a side mounted machine gun, fired from an aircraft in a constant radius turn, to be used for anti-submarine work. The idea was lost in the massive arming and subsequent de-arming of American forces. But in the early 1960s, when MacDonald, then at Eglin Air Force Base, met Ralph Flexman, an engineer with Bell Aerosystems. Flexman had read about an American missionary in South America who had delivered mail and supplies to remote villages by lowering the goods at the end of a rope and flying constant radius turns around it thereby keeping it relatively stationary while villagers removed goods and reloaded it with outgoing mail.

If a direct line – a rope – to the ground could be maintained by a flying aircraft, why couldn't a line of fire?

Gradually finding supporters, Flexman finally had a live fire demonstration in 1964 at Eglin. The results amazed even the staunchest supporters. The targets had literally been decimated by the fire.

Nicknamed "Spooky," the first U.S. Air Force gunship designated the AC-47 went to war in Vietnam armed with three .50-cal machine guns, but they were quickly replace with three 7.62-mm electrically fired Gatling guns.

Incredibly simple, with no sensors, fire control computers – just a gunsight mounted next to the pilot's left shoulder, Spooky was devastatingly effective.

Immediately deciding that more was better, the Air Force launched the AC-119 Shadow, now with four mini-guns and a powerful searchlight for use at night. The AC-119K Stinger followed shortly after that, adding elaborate sensors and two 20-mm cannons.

A time exposure of what a single AC-47 can do to a ground target. (USAF Photo)

The Spectre, an AC-130E, in action at dusk in Vietnam. (USAF Photo)

The newest gunship, the AC-130U, departs Edwards AFB, Calif., on a test mission. (USAF photo)

By the time the war ended, the gunship had found a permanent home in the C-130 Hercules. Now designated the AC-130 Spectre, this gunship is the biggest and most advanced gunship in the world

Today's gunship carries a host sensors, including infrared, low-level television and a system that tracks electronic signals. But the mission is still the same – put rounds on target. To do that, the Spectre is armed to the teeth 7.62-mm miniguns, 20- and 40-mm cannons, and a 105-mm howitzer that lobs 42-pound shells with unerring accuracy. With the modern electronics on board, the 14-man crew can attack two separate targets simultaneously.

AC-47 with three 7.62-mm miniguns. The miniguns quickly replaced the original three .50-cal machine guns, greatly increasing firepower. (USAF Photo)

Used almost exclusively for close air support, from Vietnam to today, the gunships have become a vital component of the battle field mix. The history of the gunship is short and sweet. They are:

Douglas AC-47 Spooky/Puff/Dragonship (acquired nicknames, no formal one given) (USA)

Simple, effective and deadly, the AC-47 was armed with three side firing, electrically operated GAU-2/A Gatling-type six-barreled 7.62-mm miniguns, and had to drop flares to see the enemy at night. The only computer it had was in its crewmen's heads, but it was powerfully effective. Its drawbacks were its low technology sensors (eyes and flares), limited loitering capability, and limited firepower (as if 18,000 rounds a minute isn't enough). The guns were aimed by the pilot through a modified camera view finders. The pilot put the aircraft into a bank, looks through the range finder, puts the crosshair on the target, and maneuvers the aircraft to keep the crosshairs on the target. The trigger is mounted on the steering column and is fired by the pilot. The trigger is electronically linked to motors on the guns in the rear of the aircraft.

The AC-47 (then called the FC-47) first saw testing action on Dec. 14, 1964, flying several daylight missions. Its first night mission came on Dec. 23. The first operational squadron began operations in early 1969.

The gunship's operations were basically the same in day or night—roll in on the target and start firing. It was pretty low tech. The fire control was all in the pilot's mind, the flares

An AC-130A at the USAF Museum in Dayton, Ohio. (Vadnais Photo)

were manually launched, and the guns were reloaded by hand.

Fairchild AC-119G Shadow (USA)

Field commanders soon called for more firepower than the AC-47 could carry. They began simultaneous tests of the C-130A and the C-119G. Everybody agreed that the C-130 was a better platform, but they were in so much demand, that no significant number could be spared. The brass dubbed the C-130 Gunship II, but went ahead and began full scale testing and development of the Gunship III, the C-119, even though it would be replaced by the C-130 when enough aircraft were available (how's that for doing things backwards).

The C-119G Flying Boxcar soon began the transition from trash hauler to night hunter. Within a few months, the powers saw that the C-119K, with the Jet Assisted Take Off package, was probably a better package, so they began work on that variant also.

The Shadow added a fourth gun and had a longer loiter time. But still not satisfied, the Air Force kept adding equipment and guns. And soon the AC-119K was born.

Fairchild AC-119K Stinger (USA)

Nicknamed the Stinger, this gunship had the four GAU-2B/A 7.62-mm Gatling guns, and added two 20-mm cannons and a searchlight to the deadly mix. It was also the first gunship to use onboard sensors to assist in finding and targeting the enemy. The biggest difference however, was that the K model had an additional two jet engines strapped on the wings that resulted in a 25 percent increase in its gross take off weight — more bullets, more flares, more fuel.

The Stinger began operational testing in Feb. 1970. The mission involved three AC-119Ks flying armed reconnaissance missions.

One of the most memorable Stinger flights occurred on May 8, 1970. During a night mission near Ban Ban, Laos, an AC-119K attacked a truck convoy. During the attack, at least

The interior of an AC-47, looking towards the tail of the aircraft. (USAF Photo)

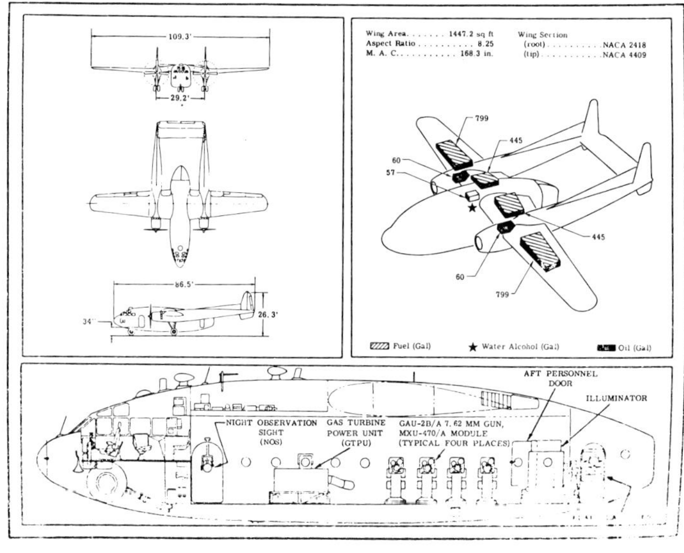

The line drawing of the AC-119 gunship. (USAF drawing)

A 105-mm howitzer being installed in an AC-130E as part of the Pave Aegis weapons upgrade program. (USAF Photo)

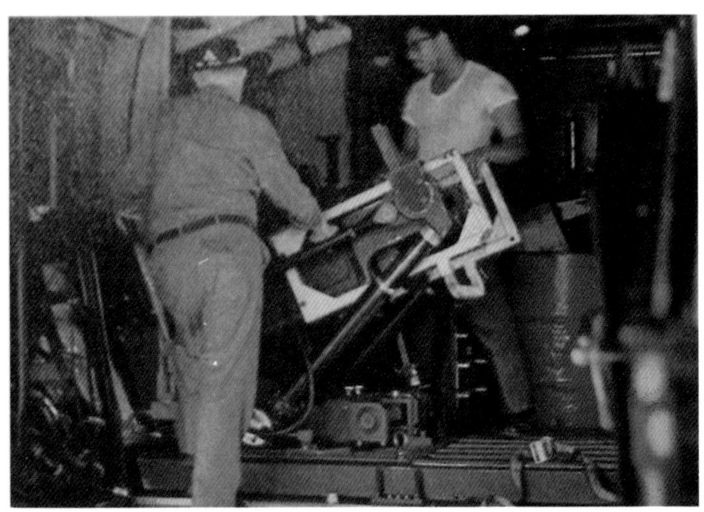

The howitzer in place. (USAF Photo)

six AA sites opened fire on the aircraft. Enemy rounds struck the right wing. A sickening dive dropped the Stinger 1,000 feet in just a few seconds. Only by using the full left rudder and full left aileron was the pilot able to regain control. Nursing the crippled aircraft back to Ubon Royal Thay AFB, the crew landed and discovered about 14 feet of the right wing missing.

The Stinger proved to be a worthy successor to the AC-47 for the short time they were fielded. Soon, attention and money focused on the ultimate gunship, the AC-130.

Lockheed AC-130A Spectre (USA)

The biggest and most advanced of the Vietnam era gunships, and still the basic platform used for today's gunships, the Spectre went into battle in 1967. Within three years, the single squadron of AC-130s at Ubon, Thailand had destroyed more than 5,000 heavy trucks (GAU-28/A) carrying men and material to the front.

Now armed with four 7.62-mm miniguns and four M-61 20-mm cannons, the Spectre was capable of firing 34,000 rounds per minute. The Spectre also came with 5,000 lbs of armor plating to protect the crew and vital parts of the aircraft.

The sensors and fire control system were cutting edge for the time—including infrared and side looking radar sensors and a fully automatic, computerized fire control system. The C-130 was originally called the Gunboat, but that changed to Gunship II, to better explain its follow on to the AC-47 mission. The AC-130 arrived in Southeast Asia in Sept. 1967 for a 60-90 day combat evaluation. By Dec. 1, 1967, the conclusion of the testing phase, the results were already clear. The AC-130 was at least three times more effective, according to the evaluators.

In Feb. 1968, after a rapid refurbishment, the AC-130A began operation missions. It did not take long for creative minds to think, "We can do this better..."

Lockheed AC-130E Spectre (USA)

In July 1970, the Air Force gave the go-ahead to develop the AC-130E. Initial plans had the new Spectre outfitted with a flare launcher, laser target designator, low-light-level television, an electronic counter measures suite, two 7.622-mm miniguns, two 20-mm and two 40-mm guns, among a host of other improvements to upgrade both the capability and the survivability of the Gunship.

Further testing at Wright Patterson AFB, Ohio, in 1971 led engineers to add a large caliber gun to the gunship while removing one of the 40-mm guns. With the addition of a 105-mm howitzer, the final arms suite for the AC-130E was one 40-mm gun, two 20-mm guns and 105-mm recoilless rifle.

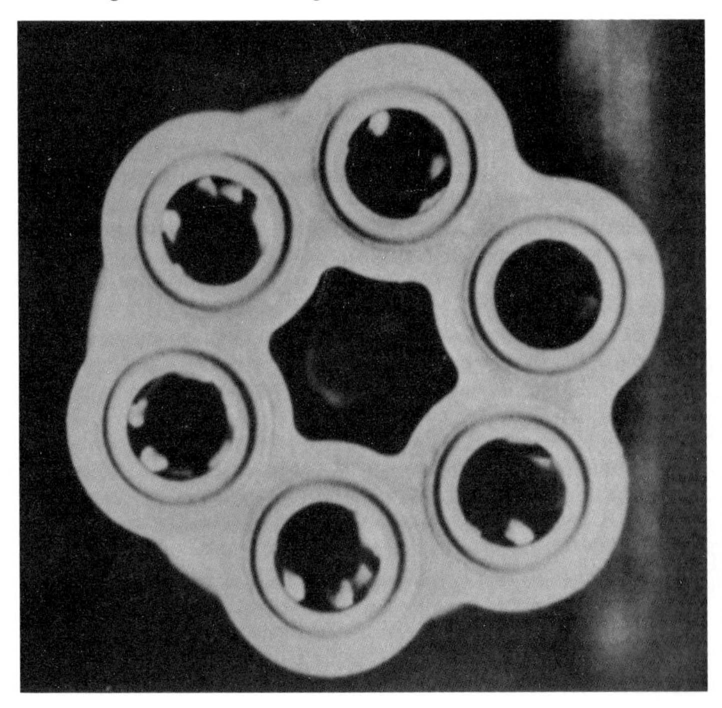

The business end of a 30-mm Gatling gun on an AC-130. (USAF Photo)

AC-130U on takeoff. (USAF Photo)

The two 7.62-mm miniguns having fallen off the drawing board some time earlier.

Lockheed AC-130H (USA)

The H model gunship was the standard bearer for the U.S. Air Force for a number of years. First deployed in 1972, the AC-130H saw action in Operation Urgen Fury in Grenada, Operation Just Cause in Panama, Operation Desert Storm in Kuwait and Iraq, and Operations Continue Hope and United Shield in Somalia. They are on duty today in Bosnia, along with its newer version, the AC-130U. The H model has a 20+ year legacy of excellence in close air support and armed reconnaissance.

The AC-130H has the same armament as the AC-130E, but has uprated engines for a higher gross take off weight and longer loiter time.

Lockheed AC-130U Spectre (USA)

Made its inaugural flight on Dec. 21, 1990, from Rockwell's facility at Palmdale, Calif., to Edwards AFB, Calif., about 2.5 hours away. This was the first of 12 new gunships bought under a 1987 contract worth nearly $800 million. This upgraded Spectre has advanced infrared and electro optical sensors, giving it the ability to track and fire on two targets simultaneously. Armed with its 105-mm, 40-mm and 25-mm guns, the AC-130Us are based at Hurlburt Field, Eglin AFB, FL.

The AC-130U is one of the most complex aircraft weapon systems in the world today. Its 600,000 lines of computer code incorporates all-light television, infrared detection system and a multi-mode strike radar. It also is pressurized, enabling it to fly at higher altitudes which saves fuel and time, allowing for a greater range or more time on target than the AC-130H. This newest version of the Gunship is seeing action in the Bosnian theater of operations.

Helicopters

The early days of military use of the helicopter were strikingly similar to the first days of the biplane. They were used for noncombatant uses — primarily transport and rescue.

But the potential for taking the war to the enemy was soon recognized and the first flight of an armed helicopter

AC-130 Gunship, with the Pace Crow sensor system protruding from the fuselage just behind the nose, being readied for deployment to Southeast Asia. (USAF Photo)

took place in 1942 at Wright Field, Ohio. There, a Sikorsky XR-4 was adapted for bombing tests.

It was not a high-tech kind of test, as the passenger heaved a 25-pound bomb over the side at the target. The early successes quickly resulted in a more traditional and sophisticated style of arming — bomb racks, sighting devices and the like.

While the results were promising, many difficulties needed to be overcome before helicopters could play a leading role in firepower. Among the early shortfalls were limited payloads and lack of overall speed.

One area that did lend itself to the capabilities and limitations of the helicopter was antisubmarine work. The slow speed of the helicopter was still faster than the fastest submarine, and the ability to land and take off vertically meant that even the smallest Navy ship could be a 'base' for ASW work.

The first ASW application occurred in 1950 on a Sikorsky S-55. Because of the limited payload, the helicopters were paired into hunter/killer teams, with one bird having the sensors to find the enemy sub and other carrying the big stick to kill the sub.

Meanwhile, Marine Experimental Squadron HMX-1 began testing in 1949 in equipping helicopters with a variety of weapons, ranging from bombs to mortars. Again, due the limited speed and payload, the school of thought was that helicopters were better suited to transportation than fighting.

The Army came onboard in 1956 when a small group of volunteer pilots gathered at Fort Rucker, Ala., to flight test most of the helicopters then in service with a variety of weapons. The tests included flying and firing weapons from .30-cal machine guns to 80-mm rockets, recoilless rifles to chemical bombs.

In 1962, Lt. Gen. Hamilton H. Howze chaired a board called the Army Tactical Mobility Requirements Board, also called the Howze Board. The board recommended moving ahead with the armed helicopter, and basing it use on a mobility concept of air assault division — the air cavalry.

UH-1Bs supporting ground troops in a rice paddy in 1965. By this time, the gunship had acquired the nickname "Cobra." (US Army Photo)

Vietnam became the proving ground for both the armed helicopter and the tactics developed to use them. The UH-1 Iroquois in theater by the end of 1967 eventually numbered nearly 3,000. They performed a variety of missions from rescue and transportation to, yes, gunships.

In the early days, the gunships were relatively lightly armed, with two 7.62-mm M-60 machine guns on outriggers on each side of the 'copter. They were aimed by the co-pilot via a reflector-type gunsight. The guns were belt-fed from ammo boxes stored in the helicopter cabin. The armed Iroquois of Vietnam became known as "Cobra's" for their deadly bite.

Other versions of the UH-1 were modified to carry rocket pods, 40-mm grenade launchers, TOW missiles, 7.62-mm miniguns, 2.75-in rockets, mortars, and Agent Orange dispensers.

A faster, better armed version of the UH-1 developed by Bell Helicopters, called the AH-1 Hueycobra, was selected over versions from Sikorsky, Kaman, and a highly modified Chinook CH-47 for gunship-only purposes. All subsequent gunships trace their lineage directly to the AH-1, the successor of the highly successful UH-1 Cobra.

Since the AH-1 arrived during Vietnam, only one other attack helicopter has seen service in U.S. colors. The McDonnell Douglas built AH-64 Apache. Though it didn't see combat until Desert Storm, once in combat it was a star. As a matter of fact, it was the opening act of the war. A deep penetration strike by the Apache gunships targeting air defense sites and command and control centers at 2:38 a.m., Jan 17, 1991, started the coalition air campaign.

In this age of high tech, the lowly chopper gunship fired the first rounds. Why? Simple, it could get to the target unseen by flying nap-of-the-earth and kill the target before the target knew what was going on. This early morning strike opened the air corridors that allowed the fixed wing guys to put bombs on target with devastating results.

The next generation gunship is about to fly—the RAH-66 Comanche. It combines the high tech stealth technologies with war tested weapons. The Comanche will carry 30-mm chain gun as well as a number of other weapons. Its advanced sensors will enable the pilot and gunner to fly, fight and win in any weather condition, day or night.

But is the Comanche the ultimate helicopter gunship? Many think that the evolution will continue and reach a point where a fighter-helicopter exists. The current generations of gunships, and the next generation under development, are all multi-mission helicopters. They have some air-to-air capability, but mostly they are built to support ground combat. A fairly lightweight, single purpose helicopter built to hunt and kill other helicopters, isn't a radical idea. Back in WW II, the Spitfire was designed and built for one thing and one thing only – air-to-air combat. It wasn't to carry bombs, rockets or strafe the ground, it was built to shot down Nazi airplanes. And was it effective!

Where will the helicopter gunship end up? How knows, but one thing is for sure, the expectations for helicopters is growing every day!

Bell UH-1 Cobra (USA)
The Huey is to helicopters as air is to breathing. Without the Huey, none of the following helicopters would exist as they do today. More than a thousand of these birds are still flying, out of God knows how many thousands were made. They first saw action in Vietnam as "slicks," unarmed troop carriers. That didn't last long, as the instinct for self preservation kicked in, and the aircrews began arming them with darn near every weapon made. From mere .30-cal machine guns, to rockets and horowitzers, from smoke generators to chemicals, the Huey carried it all into battle.

It would be impossible to list all the different weapons carried on the Huey – it would easiest to say that if it was made, the Huey flew it.

Cobra's at the Bell manufacturing plant, being readied for delivery to U.S. Army pilots. (US Army Photo)

Bell AH-1 Hueycobra (USA)
The lesser known of the two premier U.S. attack helicopters in the world. Now called the Cobra, it is the choice of helicopter for the U.S. Marines, and several other countries, including Turkey, Taiwan and Sweden, among others.

The Cobra is a derivative of the UH-1 Huey, basically a narrowed body with the crew sitting in-line rather than side by side. Introduced to action in Vietnam, the Cobra was a lethal escort to the slicks and the unslick.

The Cobra now packs a three barrel M197 20-mm gun with 750 rounds. Its four hardpoints can carry a variety of weapons from TOW anti-tank missiles to 2.75-in rockets.

McDonnell Douglas AH-64 Apache (USA)
The awesome AH-64 was born in the mid 1970s, and two decades later, it remains a substantial player with the U.S. Army.

This gunship is armed to the hilt with side pylons that carry Hellfire anti-tank missiles and 2.75-in rockets. But the interesting aspect of the firepower is the M230 30-mm Chain Gun. It's not the fact that the gun is carried on the chopper, it's how it's carried.

It's mounted on the fuselage, but it' not a clean installation with modern fighters. Since aerodynamics don't come into play in the application, the gun is slung out into the airstream directly below the canopy. Quite frankly, it looks a bit like an afterthought, but it definitely can get the job done.

Sikorsky S-70/UH-60/MH-60 Seahawk/Blackhawk/Pave Hawk (USA)
It's known by a number of different names, but for he most part, it's called the Blackhawk. Like other offensive choppers, the Blackhawk uses a montage of weapons mounted pylons including Hellfire ASMs, rockets, mine dispensers, jamming flares an chaff dispensers.

The guns are there, but not on the pylons, but like earlier gunships, are mounted din outward-firing positions in the sides of the fuselage. Depending on the particular configuration, there are either one or two 7.62-mm M60 machine guns.

McDonnell Douglas OH-58 Kiowa/Kiowa Warrior (USA)
More than 2,000 OH-58A began service with the Army in 1970s as scout and liaison helicopters. About half were later upgraded to OH-58C standards with a more powerful engine and new avionics. But their mission didn't evolve until the early 1990s.

Beginning in 1994, nearly 300 Kiowas are being converted to Kiowa Warriors, armed scout helicopters.

Armed with .50-cal machine guns, Stinger missiles and 2.75-in rockets, this is one scout that packs a punch.

Eurocopter AS 365 Panther/Dauphin (German-French)
These two helicopters share a similar platform, but different functions. By and large, the Dauphin is decked out as a civilian search and rescue or ambulance helicopter while the Panther has a more sinister role.

Armed with four anti-ship missiles—but no gun—and various self protection devices, the Panther serves largely in the Middle East. Saudi Arabia and United Arab Emirates are big users of the Panther.

The U.S. Coast Guard uses Dauphins for search and rescue operations.

The prototype YUH-1D was fitted with the M-6 armament subsystem on an experimental basis (Bell Photo)

An AH-1G Cobra gunship comes in for a landing after a training mission in Vung Tua, Vietnam. (US Army Photo)

Eurocopter AS532 Cougar (German-French)
The Cougar is a militarized version of the Super Puma. The Cougar comes in two basic variations, one for naval warfare and one for land.

The naval version carries anti-ship missiles, sonobouys, or light weight torpedoes for anti-submarine work. They also come with a folding tail and a few other navy-unique traits.

The Cougar sees action in Kuwaiti, Jordanian, Saudi, and Omani armed forces.

Eurocoper AS 350/355 Ecureuil (German-French)
This nifty little scout/gunship is armed with a single 20-mm gun, twin 7.62-mm gun pods, TOW anti-tank missiles or folding fins rockets.

Mil Mi-24/25/35 Hind (Russia)
Probably the most famous Soviet helicopter, the Hind is a heavily armed, very tough gunship. It was a feared weapon in Angola before the U.S. backed Afghans began using Stinger missiles to fight them.

Literally armed to the teeth, the Hind carries one 12.7-mm Gatling style gun in the nose, four underwing hardpoints that can carry any combination of 57-mm rockets, 80-mm rockets, 23-mm gun pods, 7.62-mm gun pods, 30-mm grenade launcher, or more than 3,000 pounds of bombs or mine dispenser. No wonder the Hind has such a fearsome reputation!

Mil Mi-8/Mi-17 Hip
The Mi-8 and its improved version, the Mi-17 are the standard armed transport helicopter for a number of former Soviet-sponsored countries. Algeria, Egypt, Libya, Iraq, Syria and Yemen fly the Hip.

For a troop carrier, it is heavily armed, with a 12.7-mm machine gun up front, and 64 rounds of 57-mm rockets as standard. In its various configurations, it can carry up to 96 rockets or four anti-tank rockets

Mil Mi-14 Haze
Another variation of the Hip, this amphibious helicopter is used primarily for anti-submarine work by Libya and Syria. It can carry torpedoes, depth charges, bombs as well as retractable sonar, sonobouys, or a Magnetic Anomaly Detector array.

The AH-1G HueyCobra attack helicopter. (US Army Photo)

Aerospatiale SA 330 Puma (British-French)
The Puma is a British-French effort to built a multi mission helicopter. It can carry up to 16 troops, litters, or 7,000 pounds of cargo. Weapons are an option on the Puma, with the basic configuration including two side mounted 23-mm gun pods, four anti-tank missiles or rocket pods, and 12.7-mm machine guns mounted in each door.

Aerospatiale SA 342 Gazelle (British-French)
Again, a British-French built helicopter, the Gazelle is used primarily by Middle Eastern countries, including Egypt, Syria, Kuwait, Morocco, and Libya.

The Gazelle is primarily an armed escort, carrying a crew of two and up to three passengers. The important passengers are mounted on outriggers, and include six anti-tank missiles, 68-mm or 2.75-in rockets, two 7.62-mm machine gun or one 20-mm cannon

Westland AS-61 Sea King (Britain)
Primarily a troop carrier or VIP transport, these Westland built, Sikorsky designed birds fly mostly in the Middle East.

Up to 28 passengers (or one VIP) can be carried up to 250 miles. The Sea King has built in provisions for weapons, and can carry a mixture of guns, rockets, missiles, torpedoes, mines, depth charges, depending on the mission need.

Kaman SH-2 Seasprite (USA)
Flown by the U.S. Navy as a submarine hunter, a number were also sold to Egypt after the Gulf War. The Seasprite carries the Light Airborne Multipurpose System (LAMPS) for hunting subs, and also carries two torpedoes for killing any sub found. When it runs out of torpedoes, it can drop one of its eight smoke markers for others to shoot at. It also carries can carry two 7.62-mm machine guns in the doorways, but usually flies without them.

The Future
Three prime contenders are vying for the helicopter of the future. Two are from the traditional bastions of helicopters — the U.S. and Europe, while the third is from upstart weapons exporter South Africa.

Denel Rooivalk (South Africa)
Currently in production, the Rooivalk is manufactured by Denel Aviation, South Africa and have sold 16 to the South African Air Force. Its design looks traditional, in the vein of the AH-64. It's wide, slab sided, and uncovered rotor head limits its stealthiness. The manufacturers claim that it has a low radar cross section, visual, infrared and acoustic signature. Be that as it may, it makes up for any deficiency in firepower. It is twin engined, armed with a 20-mm Armscor cannon, with 700 rounds of ammunition. The chin mounted gun can slew 90 degrees in just one second. The Rooivalk also carries and electronic warfare suite, eight or 16 Mokopa anti-tank missiles, Mistral air-to-air missiles, or either 38 or 76 68-mm rockets. The Rooivalk is being marketed to countries in the Middle East and Southeast Asia.

Kamov Ka-50 Hokum (Russia)
The world's only single seat close support helicopter, the Hokum loads the pilot/gunner with every task associated with flying and fighting. The Russian Aviation branch has ordered the Hokum, and they should be entering active service soon. A distinguishing mark of the Hokum is the twin, counter rotating rotor heads. This duel rotor head configuration eliminates the need for a tail fan. It also has an ejection seat system that once the handle is pulled, explosives blow the rotor heads off and then the seat kicks out.

The Hokum carries a huge amount of weapons for such a small helicopter. Along with its single barrel, 30-mm 2A42 cannon (280 rounds), it is armed with up to 80 80-mm air-to-surface rockets or up to 12 laser guided anti-tank missiles. It has provisions for additional 23-mm gun pods, R-60 'Aphid' or R-73 'Archer' air-to-air missiles, anti-radiation missiles, 500 pounds of bombs or other dispenser munitions.

Eurocopter Tiger (German-French)
A step into the future, but not a leap, the Tiger uses advanced electronics and flight control systems to make a very agile and powerful gunship, but it too has limited stealthiness. Armed with a 30-mm cannon. It also carries a wide assortment of rockets, missiles, and guns.

Boeing RAH-66 Comanche
Built by Boeing, the Comanche is the next generation of light attack helicopter. It incorporates stealth characteristics, weapons stored internally (except for the nose gun) and advanced sensors, including chin-mounted video camera and infrared sensors. It can carry up to 14 Hellfire air-to-ground missiles or 18 Stinger air-to-air missiles, and its 20-mm nose gun can rotate 360 degrees for both maximum targeting and to reduce drag in speed runs. The Comanche also includes an automatic Target Detection/Classification computer system to help the gunner identify and classify targets.

Appendix: William Tell

In 1954, the U.S. Air Force started William Tell, the USAF Air-to-Air Weapons Meet. This competition began as the air-to-air rocketry portion of the Third Annual USAF Fighter Gunnery and Weapons Meet. These early meets had teams from Air Training Command and Air Defense Command competing for top honors.

In 1958 the competition moved from Yuma, Ariz., to Tyndall Air Force Base, Fla., and has remained there since. Additionally, the 1958 meet saw the first demise of the machine gun/cannon portion of the meet as the Air Force focused on stand off weapons like missiles and rockets.

By the 1961 meet, all fighters competing were supersonic, carried missiles as their primary weapon and focused on air defense scenarios rather than pure air combat. It wasn't until 1982 that the name, and the focus, changed to the USAF Air-to-Air Weapons Meet. This memorable year also saw the F-15 Eagle compete for the first time. More strikingly, the meet reintroduced the air-to-air gunnery competition with two categories, the 20-mm cannon category for the F-15s and the Gun category for the gun-equipped F-106s. The F-15 won the overall that year as well.

After being canceled during the Gulf War, William Tell 1992 was the first meet to be sponsored by the newly created Air Combat Command. This meet also saw the introduction of the F-16 to the competition.

What began as a mere air-to-air rocketry contest has evolved into a competition that benefits the entire Air Force. Testing the mettle of aircrew, weapons controllers, aircraft maintenance and weapons loading specialists, William Tell now measures the entire spectrum of basic combat skills needed to fight air-to-air.

The Legend

The name William Tell, though associated with the Air Force's fighter interceptor meet, actually dates back to 1307 when Austria occupied Switzerland.

The Swiss overseer in the canton of Uri was named Gessler. Gessler was pretty much despised by everybody, mostly because of his pig-headed arrogance. One of his edicts was that everybody had to bow down to his hat. This hat, placed on a post in the village square, symbolized the Austrian sovereignty over the Swiss, and Gessler's disdain for the Swiss people. The penalty for failing to bow was death. Legend has it that a Swiss peasant, William Tell, traveling with his young son, refused to bow.

Because of this failure to bow, Tell was given three choices: Before the whole village, bow to the hat; die at the chopping block; of fulfill Governor Gessler's cruel taunt of proving his renowned marksmanship by shooting an apple off his own son's head. With this last choice came freedom for William Tell and his son.

There was only one choice for the pair, William and his son would never give in to a foreign bully. Taking the offered apple from Gessler, Tell's young son said that he was not afraid. The boy took the apple to far end of the meadow, leaned against a tree and gently placed th apple on his head. The father took up his cross bow, planted his feet and let fly a bolt. The crowd cheered as the apple split into pieces, leaving the boy untouched. The two rushed to each other.

Gessler was not touched by the tender scene. As Tell senior dropped the cross bow to hold his son, a second arrow fell from his sleeve. Gessler, upon seeing the bolt fall, said, "Small need for a second arrow, had your first missed." Tell's answer was, "Had my first bolt missed its mark, that second bolt was for your black heart."

Gessler, enraged, threw Tell into jail. Tell soon escaped and later killed Gessler during the fight for Swiss independence.

The legend attests for one man's willingness to pay any price for freedom, and this spirit continues in the fighter interceptor meet.

The Winners

Year	Unit	Aircraft
1954	3550th Flying Training Wing, Moody AFB, Ga.	F-94C
1955	26th Air Division, Duluth, Minn.	F-94C
1956	94th Fighter Interceptor Squadron, Selfridge AFB, Mich.	F-86D
1958	465th Fighter Interceptor Squadron, Griffiss AFB, NY	F-89J
	326th Fighter Interceptor Squadron, Griffiss AFB, NY	F-102A
	125th Fighter Group (ANG), Jacksonville, Fla.	F-86D
1959	319th Fighter Interceptor Squadron, Bunker Hill AFB, Ind.	F-102A
	460th Fighter Interceptor Squadron, Portland, Ore.	F-102A
	538th Fighter Interceptor Squadron, Larson AFB, Wash.	F-104A
1961	445th Fighter Interceptor Squadron, Wurtsmith AFB, Mich.	F-101
	59th Fighter Interceptor Squadron, Goose AB, Labrador	F-102
	456th Fighter Interceptor Squadron, Castle AFB, Calif.	F-106
1963	445th Fighter Interceptor Squadron, Wurtsmith AFB, Mich.	F-101
	146th Fighter Interceptor Squadron, Greater Pittsburgh Airport, Penn.	F-102
	318th Fighter Interceptor Squadron, McChord AFB, Wash.	F-106
1965	62nd Fighter Interceptor Squadron, K.I. Sawyer AFB, Mich.	F-101
	32nd Fighter Interceptor Squadron, Camp New Amsterdam, Netherlands	F-102
	71st Fighter Interceptor Squadron, Selfridge AFB, Mich.	F-106
	331st Fighter Interceptor Squadron, Webb AFB, Texas	F-104
1970	119th Fighter Group, Hector Field, ND	F-101
	148th Fighter Group, Duluth Int. Airport, Minn.	F-102
	71st Fighter Interceptor Squadron, Malmstrom AFB, Mont.	F-106
1972	119th Fighter Group, Hector Field, ND	F-101
	115th Fighter Group, Truax Field, Wis	F-102
	460th Fighter Interceptor Squadron, Grand Forks AFB, ND	F-106
1974	101st Fighter Interceptor Group, Bangor, Maine	F-101
	124th Fighter Interceptor Group, Boise, Idaho	F-102
	120th Fighter Interceptor Group, Great Falls Int. Airport, Mont.	F-106
1976	142 Fighter Interceptor Group, Portland Int. Airport, Ore.	F-101
	120th Fighter Interceptor Group, Great Falls Int. Airport, Mont.	F-106
1978	147th Fighter Interceptor Group, Ellington AFB, Texas	F-101
	86th Tactical Fighter Wing, Ramstein AB, Germany	F-4
	49th Fighter Interceptor Squadron, Griffiss AFB, NY	F-106
1980	147th Fighter Interceptor Group, Ellington AFB, Texas	F-101
	347th Tactical Fighter Wing, Moody AFB, Ga.	F-4
	144th Fighter Interceptor Wing, Fresno, Calif.	F-106
1982	409th Squadron, CFB Comox, CanadaC	F-101
	49th Fighter Interceptor Squadron, Griffiss AFB, NY	F-106
	57th Fighter Interceptor Squadron, Keflavik AB, Iceland	F-4
1984	33rd Tactical Fighter Wing, Eglin AFB, Fla.	F-15
	142nd Fighter Interceptor Group, Portland Int. Airport, Ore.	F-4
	177th Fighter Interceptor Group, Atlantic City, NJ	F-106
1986	33rd Tactical Fighter Wing, Eglin AFB, Fla.	F-15
	119th Fighter Interceptor Group, Fargo Int. Airport, ND	F-4
1988	49th Tactical Fighter Wing, Holloman AFB, NM	F-15
	142nd Fighter Interceptor Group, Portland, Ore.	F-4
1992	18th Wing, Kadena AB, Okinawa	F-15
	120th Fighter Interceptor Group, Great Falls Int. Airport, Mont.	F-16
1994	North Dakota ANG	F-16
1996	119th Fighter Group, 4th Wing, Canadian Forces, Cold Lake, Canada	CF-18

The Best of the best, William Tell's Top Guns

Year	Name	Aircraft
1954	Capt. Clarence W. Lewis, Jr. and 1st Lt. James R. Boone	F-94C
	3550th Flying Training Wing, Moody AFB, Ga.	
1955	Col. B.H. King and Lt F.S. Goad	F-94C
	26th Air Division, Duluth Municipal Airport, Minn.	
1956(Tie)	Col. Donald W. Graham and 1st Lt. Billy R. Thompson	F-89D
	66th Fighter Interceptor Squadron, Elmendorf Field, Alaska and	
	1st Lt. Robert B. Long	F-86D
	94th Fighter Interceptor Squadron, Selfridge AFB, Mich.	
1958	Col. Robert E. Dawson	F-86D
	125th Fighter Group, Jacksonville, Fla.	
1959	Capt. Frederick H. England, Jr.	F-102A
	460th Fighter Interceptor Squadron, Portland Int. Airport, Ore.	
1961	Lt. Col. Frank R. Jones,	F-102
	59th Fighter Interceptor Squadron, Goose AB, Labrador	
1963	Lt. Col. J.W. Rogers	F-102
	317th Fighter Interceptor Squadron, Elmendorf AFB, Alaska	
1965	No overall Top Gun award given, only category winners	
1970	Capt. James Reimers and Capt. Arthur Jacobson	F-101
	119th Fighter Group, Hector Field, ND	
1972	Capt. Lowell Butters and Capt. Douglas Danko	F-101
	425th All Weather Fighter Squadron, CFB Bagotville, Canada	
1974	Maj. Ralph D. Townsend	F-102
	190th Fighter Interceptor Squadron, Boise Air Terminal, Idaho	
1976	Maj. Bradford A. Newell and Lt. Col. Donald R. Tonole	F-101
	142nd Fighter Interceptor Group, Portland ANGB, Ore.	
1978	Capt. Earl G. Robertson and Capt. Brian J. Salmon	F-101
	22nd NORAD Region, North Bay, Canada	
1980	Capt. Maurice Udell and Maj. David S. Miller	F-101
	147th Fighter Interceptor Group, Ellington AFB, Texas	
1982	Lt. Col. Jere Wallace	F-15
	18th Tactical Fighter Wing, Kadena AB, Okinawa	
1984	Capt. Scott Turner	F-15
	32nd Tactical Fighter Squadron, Camp New Amsterdam, Netherlands	
1986	Capt. John Reed (USAF exchange pilot)	CF-18
	425th Tactical Fighter Squadron, CFB Bagotville, Canada	
1988	Capt. Teddy Varwig	F-15
	49th Tactical Fighter Wing, Holloman AFB, NM	
1992	Capt. Jeffrey Prichard	F-15
	18th Wing, Kadena AB, Okinawa	
1994	Capt. James Browne	F-15
	52nd Fighter Wing, Spangdahlem AB, Germany	
1996	Capt. Steve Nierlich (Canada)	
	4th Wing, Cold Lake, Canada	CF-18

Also from the publisher